STUDIES IN PHILOSOPHY

Edited by

Robert Nozick
Pellegrino University Professor
at Harvard University

THE EXPLANATIONIST DEFENSE
OF SCIENTIFIC REALISM

Dorit A. Ganson

Routledge
Taylor & Francis Group

NEW YORK AND LONDON

First published in 2001 by
Garland Publishing, Inc.

Published 2016 by Routledge
711 Third Avenue, New York, NY 10017, USA
2 Park Square, Milton Park, Abingdon, Oxfordshire OX14 4RN

First issued in paperback 2016

Routledge is an imprint of the Taylor and Francis Group, an informa business

Library of Congress Cataloging-in-Publication Data

Cataloging-in-Publication Data is available from the
Library of Congress

ISBN 13: 978-1-138-96936-0 (pbk)
ISBN 13: 978-0-8153-3964-9 (hbk)

The Explanationist Defense
of Scientific Realism

Contents

Preface xiii

CHAPTER ONE
Introduction 1

I. Explanationism 1

    I.i   *The Debate About Realism and the Constraints of Rationality* 1

    I.ii   *Versions of Explanationism/Realism* 5

    I.iii   *Externalist vs. Internalist Explanationist Approaches to Defending Realism* 11

II. Miller's Internalist Explanationism 14

    II.i   *Introducing Topic-Specific Truisms* 14

    IIii   *A Brief Overview of Miller on Cause, Explanation, and Confirmation* 16

    II.iii *The Role of Truisms in the Explanationist Defense of Realism* 21

    II.iv *Why Truisms are Independent Marks of Rationality* 23

    II.v   *Unfinished Business in Miller's Program* 29

CHAPTER TWO
Acausal Models of Explanation 35

I. Introduction: The Threat Acausalist Models of Explanation Pose to Contemporary Explanationism 35

II. A Brief History of Acausal Models of Explanation 38

    II.i   *Hume's Legacy and the Deductive-Nomological Model* 38

    II.ii   *From the Inductive-Statistical Model to the Statistical Relevance Approach* 41

III. The Statistical Relevance Model of Explanation 46

    III.i   *An Outline of the Model and Some Refinements* 46x

    III.ii *The Requirement of Objective Homogeneity* 50

    III.iii *Salmon's Acausal Criteria for Admissible Selection Rules and Ensuing Problems* 53

    III.iv *Difficulties in the Identification of Causal Relevance*

With Statistical Relevance 60

CHAPTER THREE
Van Fraassen's Arguments against Inference to the Best Explanation 67

I. Van Fraassen's Constructive Empiricism 67

II. Building a Case Against Explanationism: The Short
  Arguments 71
  II.i   *The Scientific Image* 71
  II.ii *Laws and Symmetry* 77

III. The Bayesian Peter Objection 82
  III.i   *Van Fraassen's Dutch Book Argument* 82
  III.ii *Flaws in the Argument* 87
  III.iii *Reconciling Explanationism with Bayes'
      Theorem* 92

CHAPTER FOUR
Van Fraassen's Dutch Books 95

I. A Philosophical Application of the Probability Calculus: Using
  Dutch Book Arguments to Derive Rationality
  Constraints 96

II. The Principle of Reflection 103

III. The Temporally Extended Principle of Reflection 110

IV. A Prohibition Against Assigning a Probability Value to some
  Special Conditional Propositions 116

CHAPTER FIVE
Varieties of Explanationism and Fine's Critique 125

I. Smart's "Wouldn't It be a Miracle?" Argument 128

II. Boyd's Arguments for Realism 133
  II.i   *Boyd's Inference to the Best Explanation* 133
  II.ii *Rival Explanans and Explanandum* 137
  II.iii *The Circularity Objection and the Realist
      Package* 146

III. Naturalized vs. Non-naturalized Realism 155

CHAPTER SIX
The Transcendental Road to Realism 161

I. Fine's Criticisms of Miller's Realism 163

II. The True Source of Unreasonable Doubt 165

III. Why Taking on Isn't Good Enough 176

IV. Salvaging Realism about Molecules 184

APPENDIX 193

I. The Realist Account of Broad Empirical Scope (and its
       Bayesian Justification) 193

II. Van Fraassen's UnBayesian Rejection of Broad Scope as an
       Epistemically Relevant Virtue 197

Works Cited 201

Index 205

to my parents, to Karen, and to Todd

Preface

The contemporary debate about scientific realism is essentially about the nature of the constraints of rationality, and where these constraints come from. The anti-realist believes that empirical adequacy, logical consistency, and probabilistic coherence exhaust these constraints, whereas the realist believes that this conception is too narrow: explanatoriness can play a special role in making beliefs rational, rationally compelling, or more likely to be true. By securing the epistemic relevance of explanatoriness, the realist tries to show that for at least some cases where unobservables are involved, we are rationally compelled to believe the best explanation, or believe that the best explanation is more likely to be true than available, incompatible alternatives (even equally empirically adequate ones). This is the basic explanationist strategy.

Two realizations of this strategy strike me as the most significant ones. The less successful approach employs the tactics of scientific naturalism and externalism, attempting to secure realism by purely aposteriori methods akin to those in the natural sciences and offering a reliabilist account of why explanatoriness is epistemically relevant (judgments of explanatoriness are informed by background theoretical considerations which are, in fact, largely true). The more promising approach, in contrast, presupposes an internalist understanding of the constraints of rationality, and a non-naturalist conception of philosophical method. We gain insights into the source of rational constraint by reflecting on the preconditions for the possibility of meaning and participation in the cooperative project of inquiry. Such preconditions include the acceptance of basic evidential principles which are required for grasp of the relevant concepts. Explanatoriness can count as an epistemic virtue when the greater explanatoriness of a theory is grounded in the theory's superior conformity to these truistic commitments.

The overarching goal of this work is to show that the explanationist defense of realism can survive the wide array of attacks that have been launched against it, and has persuasive considerations in its favor, provided we stick to the internalist, non-naturalist approach.

My time as a graduate student at Cornell University was accompanied by serious challenges, and I am most grateful to the members of the (aptly named) Sage School of Philosophy, especially Richard Miller, for continuing to support me despite these obstacles. Dick was a wonderful thesis advisor for me: a philosopher I greatly admire, whose comments on my work always led me in an interesting direction (even if I didn't always fully understand them at first hearing), and a patient, understanding mentor. I thank my committee members, Jon Jarrett and Allen Wood, for their kindness and helpful comments, and Carl Ginet and Richard Boyd for challenging questions and discussion. I also wish to thank my new colleague at Oberlin College, Martin Jones, for his astute critique of chapter three.

I could not have made it to this point without the love of my parents, the insightfulness and caring of Karen Burlew, and the tireless faith of my husband in my ability to find fulfilment in my work and life. That is why this work is dedicated to them.

D. A. G.
Oberlin College
July 2000

Now produce your explanation,
and pray make it improbable.

—*from* The Importance of Being Earnest

CHAPTER ONE
Introduction

I. EXPLANATIONISM

I.i The Debate About Realism and the Constraints of Rationality

The realist/anti-realist dispute in philosophy of science has been raging for quite some time now. There is of course no clear consensus on what all the fuss is about, but it seems fair to say that much of the debate focuses on the nature of rational constraint, and whether or not explanatoriness is somehow relevant to this issue. Differences in opinion on this matter then give rise to the familiar controversy about the status of beliefs about the unobservable: the anti-realist denying, and the realist accepting that we are compelled to believe or prefer at least some claims about the unseen world. (Given van Fraassen's admission that beliefs about unobservables can be rationally acceptable, it is no longer appropriate to characterize realism as the view that we are sometimes in a position to form such beliefs rationally.)

A popular way to speak of the disagreement about the breadth of rational constraint is to say that the two parties in question accord different degrees of significance to inference to the best explanation, or see it as playing different roles in rational opinion change, at least where beliefs about unobservables are concerned. The anti-realist wouldn't mean to harbor suspicions about our acknowledgment that everyday life creates demands for explanation, or our sense that it would be outrageous to question the legitimacy of the responses we find most satisfying in some cases. (Why does the refrigerator smell so bad? Because we forgot to throw out that potato salad that's been sitting in there for a few weeks.) Whatever broad range of types of ampliative reasoning is intended to be captured by the catch phrase, inference to the best explanation when restricted to the familiar, observable part of our surroundings is not usually the subject of vexing philosophical conflict. Tensions arise as we move beyond what can be perceived.

In this realm, it is a little easier to depict the negative, anti-realist perspective. The anti-realist contends that, whatever it is that we could mean by inference to the best explanation, we are never rationally compelled to accept the conclusion of such an inference if it concerns unobservables. There are always conflicting, equally rational alternatives, at least in principle, and complete suspension of belief about the causes of the phenomena the conclusion explains (accompanied perhaps by a belief only in the empirical adequacy of the conclusion) is always a reasonable option. Furthermore, being the result of an inference to the best explanation never serves to justify counting a belief as rational, and never is required for a belief to count as rational. A more general way of describing the anti-realist position is to say that it is fundamentally anti-explanationist: it involves the rejection of the idea that explanatoriness or explanatory superiority play a role above and beyond logical consistency and empirical adequacy in making certain kinds of beliefs rational or rationally compelling. (Note that "playing a role" here could include, for example, being the justification for counting a belief as rational, or serving as a necessary condition for rational opinion about certain subjects matters.)

This characterization of anti-realism is perhaps a bit broader than one is accustomed to seeing, but I think it is fair. The spirit of anti-realism in philosophy of science really does seem to originate in the feeling that rationality is quite flexible. Logical consistency, empirical adequacy, and perhaps probabilistic coherence are virtues a set of beliefs or theory can have which are clearly relevant to rationality and getting things right or nearly right in so far as any set which hopelessly failed to have these virtues couldn't possibly be reasonable or near the mark, but, for an anti-realist, there is not much more to say on the subject. Something like an argument from queerness can fuel such a feeling. If there were more to rationality, what on earth would it look like, and where on earth would it come from? The lure of inference to the best explanation is apparently a misleading seduction. After all, the anti-realist implores, the kinds of features we typically use to rate explanations (such as simplicity) cater to purely pragmatic considerations, and seem to be completely irrelevant to reasonableness or likelihood of truth. The authority of any demand for an explanation, and any response which most appropriately satisfies that demand at a stage in inquiry is grounded in pragmatic interests, interests stemming from our particular circumstances and desires, not interests we possess purely in virtue of being rational inquirers. (I make this distinction to clarify what is intended by the contrast drawn between what we refer to as merely practical rationality and, for lack of a better word, epistemic rationality.)

I strive here to retain the full force of the term "rationality," a mysterious yet meaningful word that has been battered about in traditional philosophy for hundreds of years. Some exceedingly radical epistemological doctrine that one might be tempted to call "anti-realist" could propound that rationality itself is thoroughly and irredeemably pragmatic or culture-specific. According to this theory, any rule of rationality (including the most fundamental laws of logic) does not reflect the norms for responsible, truth-oriented inquiry that should be shared by the human community of inquirers as a whole, but rather, caters to the specific interests of particular groups (e.g. for power, for control of nature, for social harmony), interests that some inquirers in another setting, belonging to a different group could not or should not be persuaded to share.

An advocate of such a view would have no trouble conceding that explanatoriness could play a role in establishing so-called "rational constraint" for some groups, but it should be clear that this sort of admission is of no comfort to the realist and doesn't truly succeed in dissolving the distinction between realism and anti-realism as I choose to define it. The realist does not object to the idea of rules of rationality as social norms molded by the interests of investigators, but she would insist that anything that counted as a genuine rational constraint could not for its force depend on conforming to interests investigators have over and above the interests they possess in virtue of being people who seek to understand the workings of the world. If the radical doctrine were correct, it would not establish that the rules of rationality are flexible (flexible enough to include a role for explanatoriness), but rather that rationality as we know it is a sham and there are no genuine rational constraints. The radical view represents the furthest departure from realism, the most extreme, deflationary form of anti-realism. Perhaps a popular stance in the humanities today, it remains a fairly marginal trend in philosophy of science. Much of the most rigorous work in support of anti-realism in this area (as we see, for example in the writings of Bas van Fraassen) refrains from completely dismantling rationality, even if it sustains a destructive impulse.

A comprehensive overview of the positive, realist side of the debate is a bit harder to provide because there are so many ways to attempt to draw a connection between explanation and rationality, to assign a significant role to inference to the best explanation, to specify what we could mean by such an expression and when it would be appropriate or necessary to employ this kind of reasoning. What unifies these many forms of realism is the suspicion that there simply must be more to rationality than what the anti-realist would grant; there must be sources of genuine rational constraint or sanction above and beyond logical consistency, probabilistic coherence, and empirical adequacy. Part of the task of filling in the gaps left by the anti-realist would be accomplished by some sort of appeal to explanation. Nowadays, the realist position is typically explanationist: it involves the acceptance of the view that explanatoriness or explanatory superiority at times play a special role in making beliefs rational, rationally compelling, or more likely to be true. An explanationist defense of realism (as realism is most commonly understood) is, correspondingly, one which tries to secure the epistemic relevance of explanatoriness in order to demonstrate that we are rationally compelled to believe or cognitively prefer certain claims about unobservables (e.g. that molecules exist).

I.ii Versions of Explanationism/Realism

Below I list two broad classes of realist/explanationist views. In the first group, explanatory superiority bestows general rational sanction and justification, or requirement. The views in the second group all involve the idea that the rational agent must take a certain cognitive stance towards the best explanation, either in general or in particular cases. The notion of fair comparison of explanatory strength (developed in Miller's *Fact and Method*) is particularly important for this second group. Different frameworks of inquiry involve some measure of divergence in their causal repertoires and principles, and hence may suggest very different kinds of possible explanations of a given phenomenon, as well as competing verdicts on which hypothesis should count as the best explanation of the phenomenon. A justification for counting a hypothesis as best is fair to a set of contrasting frameworks of inquiry if it only makes use of claims which are permitted in all of the frameworks in the set.

Group I

1. The explanatory superiority of a belief can serve as the justification for regarding a belief as rational. If we wanted to justify counting a belief as rational, it would be enough to show how that belief can be seen as the result of an inference to the best explanation. This form of explanationism should be familiar from Russell's *Problems of Philosophy*. Our belief in the material world is rational because it is the best explanation of the play of sensations before us.

We can refine this form of explanationism somewhat by specifying that the explanatory superiority of a hypothesis, if fairly assessed relative to all current frameworks, justifies counting belief in the hypothesis as rational. This stance represents the more moderate component of the explanationist view Miller promotes: "Fair causal comparison can make preference for belief over disbelief reasonable and non-dogmatic, even though the opposite preference would be reasonable in a possible framework that has not actually arisen." (Miller 1987, p.223)

2. Explanatory superiority is required for certain kinds of belief to count as rational. Particular kinds of belief (perhaps, for example, those involving claims going beyond current sensation that do not have a criterial status) cannot count as rational unless they can be seen as involved in an inference to the best explanation. The position is stated very crudely here, but Miller explores a more sophisticated version in a draft, "In Search of Inference to the Best Explanation."

Group II

3a. It is in general true that . . .

3b. There are certain cases where . . .

. . . facing phenomena that stand in need of explanation, if there is a best explanation of the phenomena assessed and justified as best in a fashion fair to all current frameworks, a rational agent must cognitively prefer the best explanation of the phenomena over all other available, incompatible causal explanations, including equally empirically adequate ones. By "cognitively prefer [over the alternatives]" we mean "regard as more likely to be true or approximately true [than each of the alternatives]."[1] Note that the phenomenological rival is not a part of the group of alternatives we are considering here. It is not a causal explanation, and need not be understood as genuinely incompatible with the best explanation.

4a. It is in general true that . . .

4b. There are certain cases where . . .

. . . facing phenomena that stand in need of explanation, if there is a best explanation of the phenomena assessed and justified as best in a fashion fair to all current frameworks, the rational agent must accept the best explanation of the phenomena as true.

A few clarifications are in order to stave off any premature

---

1. Though I won't always clutter sentences with the qualification "true or approximately true," it should be understood as applying just about everywhere in this dissertation. In the context of the discussion of scientific theories, it is often best to speak of "approximate truth" if one is going to speak of truth at all. Even the very best theories will involve quantitative inaccuracies (however slight).

objections. All options make some room for inference to the best explanation, but none thereby presuppose that it would be possible to subsume our many instances of legitimate reasoning involving explanation under a highly general rule or set of such rules (to which we are all bound). Furthermore, despite appearances to the contrary, it is really only 4a which stands to be damaged beyond repair if we reject the idea that in cases where it is possible to assess explanatory superiority in a way that is fair to all current frameworks, there is always a conclusive, single verdict of explanatory superiority. All other forms of explanationism can, with minimal adjustment, survive the concession that there may be several independent, equally adequate, fair standards of explanatory strength which do not always work together to single out a unique best explanation. Given a tie–say between two contrasting causal accounts–believing either one becomes rationally permitted according to 1, and both satisfy the necessary condition for rationality spelled out in 2. 3b and 4b require only that there be some cases where there is no tie, and 3a is perhaps salvageable by specifying that, in cases of ties, the rational agent must believe that each of the best explanations are more likely to be true than any of the explanatorily inferior, rival causal accounts.

Strictly speaking it is only 4a & b which specify that, either in general or in particular cases, agnosticism in light of a fair inference to the best explanation is not a rational alternative. Taking agnosticism about a proposition to be simply the stance of refraining from accepting the proposition as true (or even more likely to be true than false, or highly likely), we see that, at every juncture where explanatoriness would play the role of arbiter of reasonableness, 1, 2, 3a & b allow us to suspend belief in this fashion. By requiring only that explanatory superiority serve as a justification for counting a belief as rational, 1 is the most minimal form of explanationism. It leaves open the possibility that there might be any number of conflicting but equally rational beliefs at a given stage of inquiry. Belief in the best explanation would be rational in virtue of explanatory superiority, whereas the other options would perhaps count as rational in virtue of some other feature or features they possess. 2, 3a & b do not require full scale belief in the best explanation for the relevant cases, but they do demand that we accord the best explanation a higher probability than any particular available option–a stance compatible with according it a probability less than .5. Even 2 does not force you to take on the best explanation; it just says that *if* you are going to form a belief at a certain juncture in inquiry, the best explanation is the only rational option. A consequence of this requirement seems to be that, in line with 3, you must at least regard the best explanation as more likely to be true than each alternative since these alternatives couldn't be rational choices.

4a stands as the only patently implausible alternative. Who wants to be committed to the view that, when confronting phenomena that stand in need of explanation, a rational inquirer must believe the best explanation? Awareness of severe limitations in resources, for instance, could make it perfectly reasonable to suspend judgment.[2] Otherwise, I have resisted including possible versions of explanationism which would fail to have any contemporary subscribers. For example, I do not include van Fraassen's conception from *Laws and Symmetry* of what is entailed by believing in inference to the best explanation, namely thinking that, as a general rule, the explanatory superiority of a claim should always cause you to raise the degree of credence you assign to the claim, even beyond what would be dictated by the routine application of Bayes' theorem.[3]

I believe that inference to the best explanation can play an important realist role, but also that, in some cases, a hypothesis' explanatory superiority will derive from its superior empirical strength and scope in a way that has an entirely Bayesian justification. The empirically stronger theory will win out over rivals if it continues to be empirically successful. In this case, the theory in question will have to be accorded a higher degree of credence than

---

2. For a fuller discussion of why we should sometimes refrain from believing the best explanation, see the last chapter.

3. I examine and criticize this conception of explanationism in chapter III.

the available alternative causal explanations; we shall have to regard the best explanation as more likely to be true than the current rivals. We reach our conclusion on the basis of the explanatory superiority of the winning theory, and can count our reasoning as a variety of "inference to the best explanation," but this admission by no means entails that we must grant the most explanatory theory extra weight, over and above what is dictated by the evidence, our prior probability assignments, and the probability calculus.[4]

I have also decided to exclude formulations of 3 and 4 that neglect to specify that a verdict of explanatory superiority with a justification that is *fair to all current frameworks* places certain cognitive demands on us. I suppose it is possible for someone to hold that a rational agent must always or at times believe or prefer the best explanation as evaluated by the agent's framework of inquiry, even when this choice cannot be justified in a current, rival framework. It seems to me odd to speak of rational compulsion to adopt or cognitively prefer a belief in the absence of non-dogmatic means for undermining a rival hypothesis put forward by an alternative, equally empirically adequate existing framework. Imagine, for example, two equally empirically adequate, incompatible psychoanalytic theories positing different constellations of causes responsible for various types of psychological distress. The theories, reflecting contrasting frameworks of inquiry, can be responsibly applied to particular cases, and (in the hands of competent clinicians) yield competing descriptions of the etiology of particular symptoms. Each framework provides justification for preferring a particular explanation of a symptom over the range of possibilities internal to the system (and perhaps some external to it), yet there is no justification satisfactory to both frameworks establishing that the best hypothesis advanced by one theory is better than the best hypothesis advanced by the other.

---

4. It is also worth noting that we are not necessarily concerning ourselves here with a way of evaluating explanatoriness and establishing rational constraint which is ultimately satisfactory to van Fraassen. See the appendix, "Van Fraassen's UnBayesian Rejection of Broad Scope as an Epistemically Relevant Virtue."

Both theories are empirically useful, and we sustain an interest in relieving the pain of people who are suffering, so adopting tentative belief in the best explanation as assessed in the framework of one's tradition and training may turn out to have certain practical advantages. The same holds, however, for someone working in the opposite tradition. Given the absence of an unbiased way to confront and undermine an active rival, there seem to be no grounds for saying that belief in or cognitive preference for the best explanation as decreed by the framework of one's tradition (as opposed to the best explanation as determined by a rival framework) is rationally compelled.

In our understanding of fair evaluation of explanatory superiority we presuppose, of course, that a framework that is capable of promoting a hypothesis in genuine conflict with the best explanation of a given phenomenon as assessed by our framework has to be recognizable as at least minimally rational. We wouldn't want it to turn out that the potential for rational constraint could be continuously defeated by the presence of a crazy, incoherent framework with sympathizers who reject our most fundamental, truistic causal beliefs and are immune to the force of even the most basic sort of rational argument or justification. If an argument or justification would have to satisfy even these misfits to count as fair, there could be no fair causal comparisons, no fair inferences to the best explanation. The most respectable way of ruling out the need of a justification to cater to the absurd framework is to emphasize that an absurd framework is not a framework at all, let alone a framework that can become the source of explanations which contradict the best explanation as assessed by our framework. Dispense with enough basic causal beliefs, and reject the most minimal rational constraints, and you have robbed yourself of the capacity to express a hypothesis concerning the causal structure of some part of our world.

With the exception of 4a, all our forms of explanationism really don't seem so outrageous, but they are quite difficult to defend (even in light of the fairly sketchy anti-realist criticisms spelled out earlier). There are no convenient, effortless ways of showing why they might be true, yet plenty of easy techniques for calling them into question. To decide to resist anti-realist temptations is to take on a hefty responsibility requiring great resourcefulness.

I.iii Externalist vs. Internalist Explanationist Approaches to Defending Realism

The two realist philosophers whose acceptance of this responsibility has yielded the most promising results are Boyd and Miller. Both offer innovative defenses of the realist/explanationist view that we are sometimes rationally compelled to accept the best explanation as true, even when some of its rivals are equally empirically adequate.[5] Securing the epistemic relevance of explanatoriness plays a vital role in their philosophical programs, which share the goal of demonstrating that we are rationally constrained to believe certain claims about unobservable entities in science.[6]

The common realist goals, however, cannot mask the deep, divisive differences in their explanationist approaches to defending realism. Boyd's approach is externalist and naturalist, whereas Miller's is internalist and non-naturalist, or "transcendentalist."

In accordance with his externalism, Boyd accepts the view that whether a belief counts as rational, justified, or an instance of knowledge depends on the presence of circumstances cognitively inaccessible to the agent. To be rational, for example, a belief does not necessarily have to pass standards for evaluation which the agent can assess; a belief is rational when it results from a reliable belief-forming process, even if the agent has no idea that the relevant process is in fact reliable, and even if, in principle, the agent

---

5. We can also reformulate their position into the more moderate claim that we are sometimes rationally constrained to accept the best explanation as more likely to be true than incompatible, available alternatives, even when such alternatives are equally empirically adequate. Both formulations capture the notion that explanatoriness can be relevant to our judgments of likelihood of truth, even when the explanatory superiority of a favored hypothesis is grounded in more than its greater empirical adequacy

6. Miller focuses on existence claims, whereas Boyd does not restrict himself in this way.

could never conclusively establish that the relevant process is reliable. Boyd's account of why explanatoriness is epistemically relevant is, consequently, distinctively externalist: judgments of explanatoriness are relevant to our likelihood of truth assignments over and above what the raw evidence dictates because such judgments are informed by background theoretical considerations which are, in fact, largely true.

For a philosophical naturalist like Boyd, who sees philosophy as a branch of the natural sciences proceeding by the application of methodological principles which are akin to methodological principles used in more conventional scientific investigations, this latter claim–like all other acceptable philosophical claims–must be regarded as defensible a poseriori by appeal to empirically substantiated rules of inference and justification. There are no non-deductive inference rules, methods, or claims which are acceptable a priori or posses an a priori justification.

According to Boyd, scientific realism is an empirically warranted thesis which is established by using some of the basic rules and methods (such as abduction) which are routinely appealed to in the confirmation of scientific findings. The realist, externalist philosophical package ultimately wins out over the anti-realist one because, much the same way a successful scientific theory attains its stature by unifying disparate phenomena and solving a number of different important problems at once, only the realist, externalist package allows us to address a wide range of pressing issues in epistemology, metaphysics and philosophy of language effectively.[7]

Miller, in contrast to Boyd, argues for the epistemic relevance of explanatoriness with distinctively a priori or transcendental means. His explanationism derives from reflection on the preconditions for the possibility of meaningful inquiry into a subject matter and the collective pursuit of understanding.

As an internalist, Miller wishes to locate the source of rational constraint within the cognitive grasp of the agent. What narrows

---

7. For a detailed overview of Boyd's explanationist defense of realism and the realist, externalist philosophical package, see chapter V.

down the range of legitimate, rational options among the equally empirically adequate alternatives at a given stage of inquiry is not the fact that one option best conforms to a heritage of true theoretical commitments, or reliable belief-forming practices, but rather that one option best conforms to standards for inquiry which the agent herself can see as reflecting the best cooperative strategies for trying to get at the truth–standards which are informed by a heritage of a priori, basic truistic commitments without which inquiry, and meaningful investigation would not be possible.[8]

Explanatoriness can count as an epistemic virtue over and above empirical adequacy when the greater explanatoriness of a theory is grounded in the theory's superior conformity to these truistic commitments. We are, correspondingly, rationally compelled to believe the best explanation when it is derived from such truisms. Because a host of scientific explanations which posit the existence of unobservable entities fall into this category, we are rationally constrained to believe in the existence of these particular entities.[9]

The overarching goal of my dissertation is to demonstrate that explanationism can survive the wide array of threatening criticisms launched against it, and has compelling considerations in its favor, provided that we stick to the internalist approach. I say this with the proviso that–as I discuss in chapters II and III–the externalist approach fares quite favorably against some famous criticisms, though ultimately it can be defeated (see chapter V).

The critics of realism/explanationism I thwart are the proponents of acausal models of explanation (chapter II), the constructive empiricist Bas van Fraassen (chapters III and IV), and Arthur

---

8. By stressing the role of a priori reflection on impartial, cooperative strategies for striving to understand the world (and the truistic commitments this entails), "epistemically rational" becomes–in contrast to how it is understood in externalist, naturalist accounts of rationality–an accessible, useful term of appraisal, indicative of when a belief has been arrived at in a way which is epistemically responsible, i.e. worthy of trust by others who are also engaged in the project of trying to find out about the world.

9. I provide a more detailed overview of Miller's explanationist defense of realism in the next section.

Fine (chapters V and VI), whose conception of the potential problems with explanationism point to a serious flaw in Boyd's program, but ultimately fail to undermine Miller's.

Since the internalist version of the explanationist defense of realism plays such a central role in this drama, I end the introduction with a detailed account of what Miller's view consists in, how he has tried to secure it in his writings, and which difficulties still need to be addressed. In the final chapter, in attempt to make an even stronger case for the internalist approach, I shall present some further considerations which help lend plausibility to the view, and eliminate some of the prima facie problems and loose ends Miller leaves.

II. MILLER'S INTERNALIST EXPLANATIONISM

II.i Introducing Topic-Specific Truisms

Miller expands the repertoire of rational constraint by arguing that a range of topic-specific, common-sense truisms–basic beliefs that we must have to begin to confront the world with our thoughts and questions–are independent marks of rationality. Some of these truisms serve as fundamental evidential principles, drawing causal connections between the external world and rudimentary sensory experience, or more inaccessible features of the world and the directly observable facts for which they are responsible. These truisms all take the following form: certain types of experience serve as prima facie evidence for/give us a prima facie reason to believe they are caused by certain kinds of facts in the external world. Examples Miller has used include:

- Having a uniformly colored, continuous visual impression with distinct edges in your visual field is, in the absence of any specific reason for doubt, evidence that you are looking at a material object with approximately that shape and color.
- Observing an object with sharp boundaries in a stable environment (exerting no influence) which moves around in a quirky, energetic fashion as its parts move gives you reason, in the absence of evidence to the contrary, to believe that the object you are looking at is alive.

- Observing a non-living object which moves around in a quirky, energetic way gives you reason, in the absence of evidence to the contrary, to believe that the object you are looking at is being influenced by external forces.
- If a procedure for magnification allows us to see things clearly that are visible only indistinctly without the procedure, we have reason, in the absence of evidence to the contrary, to think that the procedure is in general reliable.
- A child's crying intensely is, all else being equal, an indication that the child is in pain.

The remaining truisms are not evidential principles, but still involve fundamental causal belief. (The group should be familiar from Wittgenstein's discussion of hinge propositions in *On Certainty*.) Examples are:

- Material objects do not pop out of existence whenever they fail to be observed.
- The world did not begin when I was born.

According to Miller, the demands foisted upon us by our truistic evidential principles are quite strict. Each such truism establishes that witnessing certain observable phenomena is to be taken as not simply a reason, but a *compelling* reason for believing a particular account of what is causally responsible for the phenomena, provided there is no specific reason for doubt. If the preconditions for the applicability of a truism are met, the rational agent will have more than reason to believe the accompanying causal claim; she will have to believe the causal claim.

As independent marks of rationality, the truisms themselves are claims a rational agent must believe, provided she has had the typical sorts of experiences we all confront. Given a highly unusual turn of events (the sort of thing described in a Borges short story), a truism might lose its status as rationally compelling. Bizarre, ongoing experiences which might be interpreted as a sign of a skeptical fantasy coming true could make complete abandonment of a truism or set of truisms a rational option (even if rejection of any particular truism is never rationally required). We shall discuss this kind of defeasibility later, but for now it should be noted that abandoning a truism in this fashion is different from merely refraining to apply it in a particular situation because of the presence of a specific reason for doubt. The latter choice would be made, for example, if one had reason to think that a child crying hysterically was actually performing in a play, or reason to think that an appearance of a chessboard was actually generated by a holographic mechanism.

The topic-specific truisms play a central role in Miller's explanationist defense of realism. Before we examine this strategy, a few words about Miller's account of causation, explanation and confirmation are in order. (A thorough description of this account would be interesting, but lengthy and not necessarily helpful to developing an overview of the essentials of the positive case for explanationism which concerns us here.)

II.ii A Brief Overview of Miller on Cause, Explanation, and Confirmation

One important anti-positivist moral of *Fact and Method* is that determining what counts as a relevant cause, a superior explanation, or a case of confirmation does not proceed solely by the application of highly general, a priori principles or rules to data at hand. With the exception of the injunction that we try to avoid violating the broad constraints of consistency and empirical adequacy, adequacy rules are much more contingent and specific to a field and context of inquiry (e.g. historical stage) than the positivists have avowed. Miller stresses all this without thereby conceding that these rules are thoroughly or predominantly pragmatic and arational in the way van Fraassen would suggest.

Miller alerts us to the need to move away from positivist-style generality, and positivist fixation on consistency and fidelity to observable empirical patterns as the ultimate sources of rational constraint and understanding. Deflationary accounts of causation (in terms of statistical relevance or some related Humean offshoot), and covering law theories of explanation (deductive nomological and inductive statistical) must be replaced by more enlightened proposals. The new credo then becomes:

Cause

Our concept of cause is not finally captured or understood by an abstract definition. Its meaning is fixed by or built up from a core of elementary varieties of causes (hitting, pushing, etc.) and specific, commonplace causal principles. These principles include those which assist us in determining when a regularity requires explanation, and when it would be appropriate to add a new candidate to our list of causes (the rules for what Miller calls the "extension procedure"). Counting an event featured in the core as a case of genuine causation requires no justification; it is just part of the meaning of "cause" that these events must be seen as involving causal influence. We can defend an extension (e.g. in physics) if we have positive responses to such questions as:

Are relevant purposes served by counting the core type of events as causes similarly served by counting the new candidate as a cause?

Do our basic causal beliefs tell us that a particular regularity demands an explanation, an explanation that seems to require the introduction of a new kind of cause to our list?

Explanation

A robust notion of cause is indispensable to understanding explanation. Previous attempts to do without it by means of a covering law model of explanation are inadequate. Briefly, this failing can be illustrated by, among other things, pointing out the unattractive consequences of the model's dependence on syntactic means for differentiating adequate from inadequate explanation. (This restriction to syntactic considerations makes it impossible to rule out an explanation as inadequate on account of insufficient causal depth. We are also barred from devising a principled way of excluding superfluous elements of explanations which does not undermine the superiority of causally deep explanations in other settings.) Another method of attack is to point out specific examples of explanations which are clearly successful, yet fail to involve highly general covering laws.[10]

Pushing aside the ruins of the covering law models Miller constructs a new account based on the acceptance of the intuitive idea that a genuine explanation is an adequate description of the causes that give rise to a phenomenon. While searching for an explanation of a phenomenon we must strive to disclose causal factors present and jointly sufficient to bring about the phenomenon in question. The norms for evaluating explanatory adequacy in a particular field of inquiry at a given time are set by the investigators' conception (at that time) of what sort of features most conform to the goal of seeking out accurate causal description in the field.

There will not always be complete agreement concerning what features of explanations should be taken into consideration. Investigators working in rival, equally empirically successful traditions may have very different notions about what qualities explanations must possess to reflect our best efforts to arrive at accurate causal description. (Consider, for example, the rival psychoanalytic theories we discussed in the introduction.)

Rules for assessing explanatory adequacy are sometimes quite technical, sophisticated, and framework relative, but it is important to keep in mind that some derive from topic-specific, truistic considerations about how different causal accounts should be compared, and can involve considerable use of our most basic, commonplace causal commitments. The potential for cross-framework agreement about a verdict of explanatory adequacy or superiority at certain junctures in our investigations is still preserved.

The commonplace, truistic basis that can support such agreement is something we are rationally compelled to accept. (We shall discuss Miller's defense of this claim later.) Any norms for evaluating explanatory adequacy that possess this sort of truistic origin cannot be seen as completely arational or pragmatic. Furthermore,

---

10. For a much more detailed discussion of what is wrong with such a model of explanation, see ch. II.

the importance of the intended goal of accurate causal description in our deciding how to judge explanatory adequacy would make it quite astonishing if all our norms for appraisal turned out to be irredeemably nonepistemic and pragmatic (even if we can concede that pragmatic considerations may persuade us to work with certain norms instead of others when irresolvable framework differences arise).

Miller's sympathy with these conclusions fails to suggest, of course, that he excludes the possibility of a context dependent dimension to the evaluation of explanatory adequacy. The rules for determining whether or not a causal story counts as satisfactory are not a priori and resistant to change as time moves on. The appropriateness of a verdict of explanatory adequacy at a given time cannot be assessed without taking the historical context of the judgment into account.

There are other benign sorts of context or interest relativity which we can assume Miller would accept. (Sensitivity to some measure of context and interest relativity for explanatory adequacy is unavoidable and uncontroversial.) For example, Miller would presumably say that once interests settle what kind of explanation is sought (e.g. in the case of emotional distress, a psychological or a physiological explanation), we must try to heed the norms of searching for accurate causal description of this kind (and these are not completely arational).

Confirmation

Different frameworks of inquiry posit contrasting theoretical entities/causal mechanisms, and hence may provide competing explanations of a given phenomenon. The favored theory promoted by one framework may gain hegemony over the theories promoted by other existing frameworks for various unsavory, nonepistemic reasons (political pressure, the whims of a privileged class . . .), but this kind of case cannot count as an instance of confirmation. (Only the most cynical philosophers of science would say that such dark forces are always to blame when a scientific theory finally triumphs over its competition.)

Confirmation is only achieved with the advent of a fair means of justification. Recall that an argument which is fair to a set of frameworks of inquiry only relies upon claims which are accepted in all of the frameworks in the set. A theory (or hypothesis) is confirmed when there is an argument that is fair to all current, available frameworks of inquiry justifying the view that the theory is the best explanation (among the available alternatives) for all the relevant data that has emerged. (Miller uses somewhat different wording, stressing that the fair argument in question must establish that the theory's approximate truth "is entailed by the best causal account of the history of data-gathering and theorizing so far" (Miller 1987, p.7).)

Taken in isolation, this picture of confirmation is fairly nonpartisan, and even parallels what we might expect an historian of science to say in the absence of a philosophical agenda. Scientific theories that working scientists have counted as confirmed fit this description very well. It is only when we begin to wonder what kind of cognitive commitment should accompany a case of confirmation, or whether or not confirmation augments the rational force of a theory that we must start taking sides in the realism/antirealism debate.

If cross-framework justification for counting a theory as the best explanation is always secured by appeal to either shared, predominantly pragmatic considerations, or claims that have been adopted for predominantly pragmatic reasons (however varied these reasons may be from framework to framework), then a case of confirmation would demand no more than a recognition of the practical preferability of the confirmed theory over its current rivals, and perhaps belief in the theory's empirical adequacy (fitting the observable data reasonably well is a minimal requirement for explanatory adequacy). We would never have to think of a confirmed hypothesis as approximately true, more likely to be true than false, or more likely to be true than any of the available rivals. We wouldn't have to agree that belief in the truth of the hypothesis is more reasonable than belief in its falsehood, etc. Summing up the anti-realist perspective in its most familiar guise: we would never have to adopt even tentative belief in the existence of the unobservable entities of a theory which is regarded as confirmed.

This anti-realist stance is one Miller intends to steer away from. He acknowledges that there are cases where a theory emerges as the best explanation among the available alternatives (in accordance, I assume, with a fair, cross-framework justification), yet agnosticism and retreat to belief in usefulness is still an option. The extent to which an anti-realist would be comforted by this qualification for certain particular cases remains an open question. As we discussed earlier, agnosticism or refraining from thinking that belief in truth is more reasonable than belief in falsehood is still compatible with thinking that the best explanation is more likely to be true than available rivals (3a & 3b). In light of Miller's implicit emphasis on the epistemic character of the goals or intentions informing the development of norms for evaluating explanations, we should have our suspicions that he may believe that confirmation or a ruling of explanatory supremacy (with fair justification) in general is accompanied by a certain cognitive requirement (in keeping perhaps with 3a or 3b). At any rate, it clear that he at least accepts 4b: there are certain kinds of cases of confirmation which, for rational agents, must be accompanied by a cognitive commitment stronger than what any anti-realist would require, namely tentative belief in the confirmed hypothesis or tentative belief in the existence of the unobservable entities depicted in the confirmed theory.

II.iii The Role of Truisms in the Explanationist Defense of Realism

The explicit strategy of *Fact and Method* seems to presuppose that belief in the unobservables posited by a confirmed theory is rationally compelled when the fair, cross-framework justification for counting the theory as the best explanation only depends on beliefs which any rational inquirer must accept. If this is the proper reading of the text, the strong realist verdict is not appropriate when an explanation is considered best in virtue of specific truisms *and* additional topic-specific principles for theory assessment that are part of a shared scientific heritage, yet fail to be rationally compelled. In such a situation, we can assume that theory development could have proceeded in a radically different, though no less rational way, a way that might have allowed for a different verdict about explanatory superiority.

Unlike his predecessors, Miller does not try to establish a highly general explanationist strategy for securing "across the board" realism about unobservable entities posited by scientific theories. Rather, he insists that we look carefully at specific cases of confirmation. Sometimes we will see that the defense of a particular theory or explanation depends on complex scientific principles or assumptions which are accepted in all current frameworks, but could in principle be absent in an equally empirically successful, but non-existent framework. No strong realist conclusion can be drawn (though we leave open the possibility that a weaker explanationist stance might be fitting). Other times, we will find that the justification for accepting an explanation is so tied up with our most basic truistic beliefs that we must, on pain of irrationality, believe that the causal mechanisms appealed to in the explanation really exist.

Miller mentions several examples of unobservable entities which he thinks fall into this latter category: animalcules, genes, continental drift, electrons, and molecules. Though he does not go into the details about the truistic underpinnings of the defense of all of these unobservables, he does give us a comprehensive picture of how topic-specific truisms play a role in the justification of belief in animalcules and molecules.

Miller illustrates how topic-specific truisms are central to Leeuwenhoek's demonstration of the existence of animalcules. Leeuwenhoek, a Dutch amateur scientist who tinkered with the construction of microscopes in the later 17th century, noticed how the features of small, visible objects that he could just barely make out (such as the parts of the barely visible cheese mite), were easy to see when observed through his instruments. Details that were only vaguely suggested by unaided observation seemed to be revealed with far greater clarity and complexity.

When Leeuwenhoek then used his instruments to look at drops of water (pond water, stagnant rain water, solutions of black pepper), and saw a menagerie of strangely shaped particles moving about in a frenzy, propelled by little limbs or fins, he concluded that his instruments allowed his to see tiny creatures ("animalcules") which are plentiful in our environment, but invisible to the naked eye.

Leeuwenhoek's justification for the claim that animalcules exist involves, in addition to the "raw" empirical data (his basic observations), two common sense truisms:

1. If a procedure for magnification allows us to see things clearly that are visible only indistinctly without the procedure, we have reason, in the absence of evidence to the contrary, to think that the procedure is in general reliable.
2. Observing an object with sharp boundaries in a stable environment (exerting no influence) which moves around in a quirky, energetic fashion as its parts move gives you reason, in the absence of evidence to the contrary, to believe that the object you are looking at is alive.

Since acceptance of these truisms is rationally compelled, belief in animalcules is also rationally compelled (even though animalcules are, strictly speaking, unobservables).

Disclosing the truistic underpinnings of the defense of belief in animalcules is perhaps a bit more straightforward than doing the same for the kinds of unobservables which figure more prominently in the debates over realism, such as molecules. We shall look at Miller's efforts to show how truisms play a role in Einstein's Brownian motion defense of the molecular hypothesis in a later section.

II. iv Why Truisms are Independent Marks of Rationality

Miller's central explanationist strategy relies on the assumption that there are truistic causal principles which any rational inquirer must accept. As fruitful and intuitively attractive as this assumption may be for a realist, it still needs to be defended–a challenge which Miller has tried to meet using some interesting approaches.

One important tactic Miller uses is to draw inspiration from what could be regarded as the later Wittgenstein's attempts to respond to radical skepticism, a skepticism of the sort that denies that even our most ordinary beliefs are rational. (A radical skeptic might say, for example, "how can our belief in the external world be justified or count as rational, when we are incapable of ruling out alternative explanations for the play of sensations before us—such as the hypothesis that we are brains in vats, or are being deceived by an evil demon, etc.?")

In *On Certainty*, Wittgenstein (at least implicitly) takes issue with the idea that a belief about the world always stands in need of justification if it is to count as rational. For inquiry to be possible at all, something must hold fast; there must be a limit to the demand for justification, for otherwise we would get caught in an infinite regress, and could never even begin our investigations.

> 115. If you tried to doubt everything you would not get as far as doubting anything. The game of doubting itself presupposes certainty (Wittgenstein, *On Certainty*)

Some beliefs (such as "I have two hands," "the world did not begin with me") are so basic and central to our way of thinking, what could we possibly appeal to in the course of attempting to justify them that would strike us as more solid or compelling anyway? Could anything conclusively speak against them?

> 111. . . . I want to say: my not having been on the moon is as sure a thing for me as any grounds I could give for it.

> 125. If a blind man were to ask me "Have you got two hands?" I should not make sure by looking. If I were to have any doubt of it, then I don't know why I should trust my *eyes* by looking to find out whether I see my two hands? What is to be tested by what?

Without these central beliefs, which Wittgenstein calls "hinge propositions," there could be no game of distinguishing between the rational and the irrational, of seeking understanding, of trying to figure things out . . . Acceptance of these beliefs is just a part of following the rules for the activity of making judgments, or speaking meaningfully about the world. They are the measure, not the measured.

> 94. But I did not get my picture of the world by satisfying myself of its correctness; nor do I have it because I am satisfied of its correctness. No: it is the inherited background against which I distinguish between true and false.

> 105. All testing, all confirmation and disconfirmation of a hypothesis takes place already within a system. And this system is not a more or less arbitrary and doubtful point of departure for all our arguments: no, it belongs to the essence of what we call an argument. The system is not so much the point of departure, as the element in which arguments have their life.

Because hinge propositions play this primary role, no deeper justification is needed or required for their acceptance to be in order, that is, to be rational: to ask for a deeper justification is to misunderstand their function in our belief system, much the same way asking "Why do you need a hoop and a ball to play basketball?" betrays a fundamental misunderstanding of basketball.

> 341. That is to say, the *questions* that we raise and our *doubts* depend on the fact that some propositions are exempt from doubt, are as it were like hinges on which those turn.

> 342. That is to say, it belongs to the logic of our scientific investigations that certain things are *in deed* not doubted.

> 343. But it isn't that the situation is like this: We just *can't* investigate everything, and for that reason we are forced to rest content with assumption. If I want the door to turn, the hinges must stay put.

> 369. If I wanted to doubt whether this was my hand, how could I avoid doubting whether the word "hand" has any meaning? So that is something I seem to *know* after all.

But belief in hinge propositions is not merely rationally acceptable. If a rational belief is a belief which reflects a good strategy for trying to find out about the world, and belief in hinge propositions is required for us even to begin to speak meaningfully about or gain a cognitive grasp of the world, then the acceptance of hinge propositions is rationally required.

I think that there is nothing in Wittgenstein's story to suggest that the hinges must remain absolutely fixed as time goes on. There seems to be plenty of room for allowing for at least a weak kind of defeasibility. Perhaps no amount of strange experience would force the abandonment of any particular hinge proposition, but the situation may be suitably bizarre to require that some hinge or other be dismantled, with considerable freedom about which one or ones. Very unusual data could loosen the grip of rational constraint, so the rational requirement that you believe a particular hinge proposition can change to the mere rational acceptability of such a belief, provided changes are made elsewhere in your belief system (some hinge proposition or other must be abandoned).

By admitting this kind of weak defeasibility, we allow ourselves to draw a more perfect parallel between Wittgenstein's hinge propositions and Millers topic-specific truisms. As Miller seems to suggest in "Absolute Certainty," we are never rationally required to abandon any particular truism, no matter how odd the evidence before us (hence the appropriateness of the "absolute certainty" with which we accept such truisms); nevertheless, abandonment of the truism may become at least a reasonable option. I assume he would agree that some uncanny situations might require the rejection of some truism or other, though which one or ones is not rationally constrained.

Miller has fashioned his topic-specific truisms from Wittgenstein's hinge propositions, and so we can expect that the Wittgensteinian defense of the claim that belief in hinge propositions is rationally acceptable and required should also hold for Miller's truisms. Truisms and hinge propositions seem to be pretty much the same thing, with the proviso that Miller tends to rely most often on truisms construed as evidential principles which conform to the form outlined at the beginning of the chapter: certain types of experience serve as prima facie evidence for/give us a prima facie reason to believe they are caused by certain kinds of facts in the external world. These constitute the somewhat narrower class of hinge propositions which express Wittgensteinian criteria (rules for taking certain types of experience as prima facie evidence for a certain fact within a subject matter, where conforming to such rules is required to speak meaningfully about that fact or subject matter).

At least in *Fact and Method*, Miller seems to endorse enthusiastically the Wittgensteinian account of why we are rationally compelled to accept hinge propositions/truisms. Later work might suggest, however, that he feels that Wittgenstein's defense is somewhat flawed by too great a focus on linguistic sources of rational acceptability and constraint. Rejecting hinge propositions subverts one's projects as an inquirer, not simply because doing so involves clashing with norms for speaking meaningfully or making genuine judgments about the world or a particular subject matter. More significantly, such an abandonment violates norms for trying to learn about the world that everyone cooperatively engaged in such a project would want everyone to follow (what Miller appropriately calls the norms of "epistemic responsibility"). Without everyone sustaining a commitment to hinge propositions, the cooperative project of finding ways to learn about the world couldn't even get started.

In "The Norms of Reason," Miller presents the idea that a rational belief (about a particular subject matter) is an epistemically responsible belief, that is, a belief which conforms to the norms for cooperatively trying to get at the truth about that subject matter, norms which any rational agent would want others to follow as they formed beliefs about the subject matter if she were "epistemically dependent" upon them. The state of epistemic dependence is introduced in order to highlight how the relevant norms are norms the agent would want others to follow qua being an agent who is interested in pursuit of the truth, not qua being an agent who has particular desires, needs or beliefs which might clash with the interests or beliefs of other agents who are also trying to learn about the relevant subject matter. (Miller mentions

how the state of epistemic dependence is analogous to Rawl's original position, where agents are placed behind a "veil of ignorance"
of their own particular circumstances, needs, beliefs and biases so
that proper judgments about justice can be made.)

Miller sees the standards of reason as socially determined,
but not thereby pragmatic or nonepistemic. By de-emphasizing
linguistic sources of constraint, and identifying epistemic norms
of cooperative inquiry as the true standards of rationality, he has
tried to show how commitment to truisms is relevant to the pursuit of truth. He is then able to proceed with his realist ambi-
tions.

I myself am drawn to Miller's view: we should look to these
social, yet epistemic norms for a deeper understanding of what the
constraints of rationality are, and where they come from. I accept
(using a formulation which is somewhat different from Miller's,
but seems to express the same idea) that a rational belief (or set of
beliefs, a theory) is a belief (or theory) which reflects a good strategy for trying to get at the truth about the world/about a particular subject matter, where the standards for what counts as a good
strategy are just the standards everyone engaged in the project of
inquiry about the world/the subject matter would want everyone
to follow given their interest in trying to get at the truth (but not
any non-epistemic interest they might have). "Good strategy for
trying etc." should be understood in an internalist, not an externalist way.

To avoid confusion, we can add the qualification that a belief
or theory can only count as reflecting a good strategy (overall) if it
fails to reflect any particular bad strategies, e.g. a theory which
was logically consistent (hence reflecting a particular good strategy), but failed to be empirically adequate (hence reflecting a particular bad strategy), would have to count as reflecting a bad strategy (overall).

If we look at the constraints of rationality in this way, we can
more readily identify why the anti-realist conception of the norms
of reason seems so impoverished. There are more standards that
anyone qua inquirer with an interest in striving towards the truth

would want everyone to follow in the course of trying to learn about the world than:

- be logically and probabilistically coherent in your beliefs
- make sure the set of your beliefs doesn't conflict with the raw empirical data (i.e. secure empirical adequacy)

Some simple examples of such standards are:

- be critical and open-minded, willing to grapple with alternative viewpoints, to engage in self-criticism, self-questioning
- don't accept a belief just because it satisfies a deep psychological need (other than a need to try to find out the truth)
- find ways to test your theories (not simply against familiar experience, but also against the beliefs of others)
- (most controversial) strive to minimize explanatory loose ends, and maximize explanatory virtue. (Fulfilling this last injunction would involve, as we have learned from Miller, minimizing conflict and maximizing integration with truistic causal principles.)

A desperate anti-realist might attempt to argue that satisfying such norms is part and parcel of trying to secure empirical adequacy, but I wouldn't be convinced. Someone could insist on upholding a theory which was, strictly speaking, empirically adequate, but fail to follow each and every one of these norms. (I would be inclined to think that John Mack, the Harvard psychiatrist who believes that people have been abducted by extra-terrestrial aliens, would fall into this category. Fundamentalist religious leaders, such as Jerry Falwell, might also serve as additional examples.)

II. v Unfinished Business in Miller's Program

Though I find Miller's internalist explanationism highly compelling, there seem to me to be three major criticisms to which his proposal is vulnerable:

1. Topic-specific evidential truisms dictate that certain types of experiences are, in the absence of a specific reason for doubt, to be

taken as a reason to believe a particular causal account of what has given rise to these experiences. But what counts as a specific reason for doubt, and why–as Miller's defense of realism presupposes–is the reason for belief indicated in a truism compelling enough to demand belief in the appropriate causal account (in light of the relevant experience and absence of a specific reason for doubt)? It seems as though we are sometimes in situations where we have a reason to think a proposition is true, and no reason to think that it is false, yet we rationally refrain from believing that the proposition is true. For example, an inquirer might know that she is in a state of epistemic poverty: at present she has evidence which speaks in favor of a particular hypothesis, and no data speaks against it, but she quite reasonably refrains from believing the hypothesis since her resources are too limited for her to proceed very far in her search for specific evidence of falsehood.

2. In *Fact and Method*, Miller appeals to a Wittgensteinean defense of the claim that truisms are independent marks of rationality: rejection of the truisms robs the agent of the ability of speak meaningfully about the relevant subject matter, to cognitively engage with the world. In "The Norms of Reason," Miller supplements this justification by arguing that acceptance of the truisms is a basic requirement for anyone who wishes to engage in cooperative inquiry into the world. The norms of reason, of epistemic responsibility, are the norms of cooperative inquiry–the norms which any rational agent would want others to follow as they formed beliefs about the subject matter if she were epistemically dependent on them. By why isn't merely "taking on" the truisms, using the truisms without believing them, sufficient to satisfy the purposes (of meaningfulness, of participation in cooperative inquiry) highlighted in these defenses? If a pragmatic acceptance which falls far short of literal belief is good enough, Miller's explanationist approach to securing realism is in serious trouble.

3. We can question whether Miller's case for realism about molecules in *Fact and Method* is entirely successful. The dependence on truisms in Einstein's Brownian motion argument for the molecular hypothesis doesn't appear to be as exclusive as would be

required for a strongly realist conclusion (that is, a realism of form 4b) to follow in accordance with the basic explanationist strategy Miller outlines in the text. Since Einstein seems to make use of fundamental, yet nontruistic causal principles of physics, a line of objection becomes available to the anti-realist. Though we can't be entirely sure in the absence of a fully worked out theory, it still seems possible that equally empirically adequate, though quantitatively more complex alternatives to these principles could have been developed, alternatives which would allow for the rejection of the molecular hypothesis without compromising any truistic commitments. The resulting theory would sustain the idea that matter is continuous only at the expense of mathematical elegance and convenience.

Einstein showed, using principles from the molecular kinetic theory of gasses, that the amount of osmotic pressure exerted by very small particles suspended in a pure liquid is derivable from the assumption that the motion of the particles, the "Brownian motion," is caused by their bumping against the discreet units of matter which comprise the fluid. For example: little particles are placed in a horizontal tube, a tube which is filled with a pure liquid and divided by a membrane (permeable to the liquid, yet impermeable to the particles), and potentially interfering forces (such as electrostatic forces) are excluded. The amount of osmotic pressure exerted on the membrane by the particles can be derived if you assume that the motion of the particles results from their being bumped around by the random motion of bits of discreet matter, much smaller than the particles.

Brownian motion is the type of phenomenon that quite naturally seems to stand in need of explanation, and Miller appropriately points out the truistic origins of this intuition (we call up another truism from our original list):

- Observing a non-living object which moves around in a quirky, energetic way gives you reason, in the absence of evidence to the contrary, to believe that the object you are looking at is being influenced by external forces.

It seems, however, that truisms alone are not sufficient to ground the following chain of reasoning:

> If the fluid is not made up of molecules, there is nothing capable of moving the particles, except for the force of gravity against which there can be no insulation. They will simply fall to the bottom of the tube. By the same token, if the particles are in constant irregular horizontal motion, with a net diffusion toward the other end and consequent pressure on the piston, something in the liquid is moving the particles. And the motions of discrete parts of the liquid are the only such sources not already ruled out. (Miller (1987), p. 475)

The same holds for a later passage (in a similar vein):

> [Einstein] shows that if the motive force on the particles is due to the action of submicroscopic objects– random, independent motion of bits of matter much smaller than the Brownian particles and constituting the chemical substance in question, then an experimentalist should observe certain magnitudes, magnitudes which are highly unlikely to arise unless the motive force is of this kind. (Miller (1987), p. 477)

The molecular hypothesis allows for a very neat derivation of the appropriate quantities, but how can purely truistic considerations rule out the empirically adequate rivals which exist in principle? Such a rival, positing perhaps some kind of strange field being set up between the particles, could sustain the idea that matter is continuous, yet account for the phenomenon of Brownian motion in a way which would accommodate the measured magnitudes.

From a purely practical point of view, it would make no sense to develop this kind of alternative: the molecular hypothesis already works so well, so we needn't bother. As we know, however, these pragmatic considerations alone cannot be used to show that refusal to believe the molecular hypothesis is epistemically irrational (or, correspondingly, that belief in the molecular hypothesis is rationally compelled). Since no violation of truisms seems to be involved here, the anti-realist could object, it looks like the explanationist strategy which is highlighted in *Fact and Method* may not be of much help for establishing a strongly realist verdict in this particular case.

This is not to say, of course, that important lessons from the discussion in *Fact and Method* fail to be reflected in Einstein's confirmation of the molecular theory. Miller shows how the process of judging that the phenomenon of Brownian motion stands in need of explanation, and the process of isolating out an explanation as best involve primarily the use of truistic (and, we would want to add, nontruistic) topic-specific causal principles, not the application of highly general rules or the satisfaction of general virtues.

Furthermore, our anti-realist has only cast doubt upon the feasibility of using the basic, unmodified truism strategy of *Fact and Method* to secure a strongly realist conclusion. Some weaker realist stance, such as "the best explanation is more likely to be true than available alternatives" (form 3b), may still be appropriate, though we can worry that, at least for Einstein's Brownian motion defense, what could be considered genuine available alternatives (not just phenomenological counterparts) are explanatorily inferior simply because they fail to be as adequate empirically. I'm not sure what, in this context, would make an alternative "available," other than the fact that the alternative explanation appealed only to forces and laws already known to physics. An available alternative might then take the form "matter is continuous, yet some (one or another) force xi known to physics accounts for the behavior of the particles." The alternatives would presumably be excluded on one of two grounds:

1. They contradict the tests that are used to show that such forces are not present.
2. They issue in incorrect values for the amount of osmotic pressure exerted by the particles.

If I have not misunderstood what the genuine available alternatives consist of, and the explanatory superiority of the molecular hypothesis as against its available alternatives does rest on its greater empirical adequacy, the Brownian motion case appears to be neutral between weaker realism and anti-realism at this point in our investigations. A genuinely realist verdict involves the idea that explanatoriness can be epistemically relevant, even for cases where degree of explanatoriness involves more than degree of empirical adequacy.

These three criticisms are, in my view, serious but not insurmountable. It shall be my hope, in the final chapter, to supplement and develop Miller's internalist explanationism in a way which is adequate to overcoming these difficulties. This task shall be the last step in my efforts to show that internalist explanationism bears the greatest promise for a satisfactory realist solution to the problems which characterize the contemporary debate about scientific realism.

CHAPTER TWO

Acausal Models of Explanation

I. INTRODUCTION: THE THREAT ACAUSALIST MODELS OF EXPLANATION POSE TO CONTEMPORARY EXPLANATIONISM

Contemporary philosophers wishing to embark upon an explanationist road to realism place great importance on the idea that our cognition of the world (parts seen and unseen) is imbued with causal richness. For these realists, explanatoriness is sometimes relevant to degree of credence, even when the explanatory superiority of a favored hypothesis is grounded in more than its capacity to accommodate the observable– not because good explanations conform to some highly general, topic-neutral requirements which indicate greater likelihood of truth, but rather because our judgments of explanatory superiority are sometimes informed by appeal to a rationally required core of causal beliefs (truistic and criterial commitments in Miller's view; background theoretical commitments in Boyd's). The robust conceptions of underlying or unobservable mechanisms, and the causal principles and intuitions out of which this core is constructed can play a significant role in our determination of which among a given set of alternative explanations is best, and when they do, their contribution accounts for the epistemic relevance of explanatory superiority beyond mere predictive success.

To provide an explanation for an event is, according to this view, to attempt to describe the causal factors which are responsible for the event (at whatever level of depth is appropriate given the context). To evaluate whether an explanation is adequate is to determine how well it conforms to already accepted causal principles and intuitions in light of the available data. When the superiority of an explanation in question is grounded in its greater fit with the rationally required core of causal beliefs, it must be regarded as more likely to be true than explanatorily inferior rivals, even empirically equivalent ones.

This conception is directly challenged by influential empiricist accounts of explanation which have emerged in the last half-century, beginning with Hempel's famous deductive-nomological and inductive-statistical models, and continuing on with proposals to provide a scientifically acceptable, philosophically rigorous analysis of explanation by replacing any reference to causal notions with purely statistical concepts (noteworthy contributors to this project include Reichenbach, Good, Suppes, and Salmon(1971)). The first model equates explanation with the deduction of the explanandum from general laws and initial conditions (where what counts as a general law should be syntactically determinable, without appeal to causal considerations). The second says that we have succeeded in explaining an event when we have shown that it is to be expected with high probability, given background conditions and laws. The third, the statistical relevance models, characterize explanation as the identification of factors which affect the probability of the explanandum (usually by raising probability, though, for Salmon, not necessarily).[1]

These three acausal approaches are united by an allegiance to the Humean view that since making legitimate claims about causal mechanisms and processes as they are ordinarily understood would require an epistemic access to mysterious, unobservable powers which we do not possess, such claims should be re-interpreted in terms which are acceptable to the empiricist, making them candidates for being warrantedly accepted on the basis of experience.[2] Whether a causal story counts as a genuine and ade-

---

1. The phrase "the statistical-relevance model(s)" should be narrowly construed in this chapter. I am concerned specifically with reductive, acausal statistical relevance models, not efforts, such as Cartwright's, to spell out what sorts of general probabilistic relations hold for explanations without trying analyze the concept of cause in purely statistical terms. Her formal account presupposes the notion of cause.

2. In fact, as Boyd notes (Boyd 1985), one could say that this Humean credo serves as the sole motivation and justification for Hempel's sort of analysis in the first place.

quate explanation ought to be settled without the need to consult fuzzy, epistemically suspect causal notions and intuitions. All that truly matters in the determination of whether theory A legitimately and adequately explains some phenomenon B, is—depending on which acausal account you accept—whether A includes general laws and B is to be expected on the basis of A (with deductive certainty, or high probability), or whether A posits factors which are statistically relevant to B.

Given their empiricist origins, it should come as no surprise that the acausal models have a significant anti-realist consequence: from an epistemic standpoint, incompatible but empirically equivalent explanations A and A' of some observable phenomenon B must always be regarded as equally adequate (or inadequate, as the case may be). The epistemic norms for judging explanatory superiority or bestness can be nothing more than the norms for judging whether the explanandum is to be expected on the basis of a theory,[3] or whether and to what extent the theory describes factors which affect the probability of the explanandum, and these norms are powerless to adjudicate between empirically equivalent rivals. Since they are empirically equivalent, A and A' assign the same probability to observable B: B is just as likely or expected given A as given A', and A and A' are equally statistically relevant to B. A and A' must be regarded as equally good epistemically and hence, equally likely, though of course pragmatic features such as aesthetic appeal or simplicity might incline us towards using one rather than the other.

This anti-realist consequence highlights with particular starkness why realism is threatened by these acausal models of explanation (though, given the role robust causal considerations play for the contemporary explanationist in settling the rational acceptability and relative likelihood of competing explanations, we might have easily guessed that any proposal which analyzes and reduces such considerations away would not square well with real-

---

3. For simplicity's sake, we take for granted that the components of both A and A' are appropriately law-like.

ism). Fortunately, each of the models has fatal flaws.

After surveying the historical progression from Hume's analysis of causation, to Hempel's two accounts of explanation, to the statistical relevance approach (where each subsequent model can be seen as an attempt to overcome the limitations of its predecessor while preserving ultimate philosophical motivations), and highlighting the now familiar difficulties with Hempel's proposals, we shall turn our attention to the last link of this chain. We shall see that even a highly refined version of the statistical relevance model exhibits strain under the pressures of overwhelming technical obstacles and philosophical problems, the most significant of which is the failure of the reductive project. Rather than being reducible to purely statistical concepts, causal notions turn out to be indispensable to the resolution of problematic cases and the successful understanding and application of a key component of the model (the concept of objective homogeneity).

II. A BRIEF HISTORY OF ACAUSAL MODELS OF EXPLANATION

II.i Hume's Legacy and the Deductive-Nomological Model

The intimidating historical precedence for rejecting the attitude towards causation and explanation which Miller and Boyd sustain begins with Hume, who acknowledged the important role causal concepts play in our judgments about the world beyond immediate experience, yet famously de-mystified and deflated what those concepts could amount to.

> For surely, if there be any relation among objects, which it imports to us to know perfectly, it is that of cause and effect. On this are founded all our reasonings concerning matter of fact or existence. By means of it alone we attain any assurance concerning objects, which are removedfrom the present testimony of our memory and senses. The only immediate utility of all sciences, is to teach us, how to control and regulate future events by their causes. Our thoughts and enquiries are, therefore, every moment employed about this relation: Yet so imperfect are the ideas which we form concerning it, that it is impossible to give any just definition of cause, except what is drawn

from something extraneous and foreign to it. Similar objects are always conjoined with similar. Of this we have experience. Suitably to this experience, therefore, we may define a cause to be *an object, followed by another, and where all the objects, similar to the first, are followed by objects similar to the second.* (Hume, *An Enquiry Concerning Human Understanding*, p.51)

According to Hume, when we make a causal claim of any form (A caused B, A's cause B's, A causes B), we are really expressing the idea that there is a constant conjunction between A and B. To make the view even minimally tenable to modern minds, we add on the qualification that the constant conjunctions hold relative to certain background conditions. For example, consider the claim "the water freezing solid is caused by the water's temperature dropping below 0° C." The relevant constant conjunction rule is: under conditions of normal atmospheric pressure, water temperature dropping below 0° C is always followed by the water freezing solid.

Hume himself did not explicitly tackle the subject of scientific explanation, but we can imagine what he might have said on the matter in light of our qualification. Any response to a request for an explanation of a B-type event (taken as a request for a causal history: Why B? Because A caused B) would have to involve seeing B as subsumed under a general rule of the form: given certain conditions, A's are always followed by B's (where A is some previous type of event that has actually occurred).

The causation-anxiety which plagued Hume has continued to find expression in our time. It is quite remarkable to see how loyal some philosophers are to Hume's original misgivings and recommended solutions after all these years.

I am of the conviction that causal talk is a mare's nest of confusions, snares, and delusions. In any case, I insist that a prelude to successful hypothesis testing is a precise statement of the hypothesis in non-causal language . . . the hypothesis needs to be stated precisely either as a universal or statistical generalization, without weasel causal words. (Earman, p.104)

20th century attempts to interpret and describe the broad features of scientific explanation in a rigorous, philosophical way have frequently reflected a susceptibility to Hume's malaise. The earliest approaches, most notably Hempel's, are remarkably similar to what we imagined Hume would have adopted, had he taken an interest in developing a deflationary, empiricist theory of scientific explanation.

We noted the consequence of applying a Humean-style reduction of causality to the natural assumption that a request for an explanation can be answered by appeal to a "causal" relation obtaining between the type of event to be explained, and some previous type of event. An explanation of B must be regarded as involving the subsumption of B under a general rule of constant conjunction between some previous type of event (A) and B-type events (under certain conditions, A's are always followed by B's).

For Hempel, an explanation of an event must show how the event to be explained is to be, or would be, expected on the basis of initial conditions, and general laws connecting the initial conditions and the event to be explained. According to his proposal, the deductive-nomological model of explanation, to give an explanation of B is to provide a derivation of B from general laws and initial conditions which hold.

Many criticisms of this model have been put forward, to devastating effect. The following well-known examples illustrate the weaknesses engendered by focusing on considerations of logical form (at the expense, some realists might say, of reasoning about causes) when deciding what counts as an explanation:

1. The length of the flagpole explains the length of the shadow cast by the flagpole, and not vice versa (at least in most contexts). Either one, however, is derivable from the other given the appropriate background conditions and laws.
2. The coming of the storm is to be expected on the basis of the barometer registering a drop in atmospheric pressure and appropriate laws, but the barometer registering a drop doesn't explain the coming of the storm (again, at least in most contexts).

3. A man's failure to become pregnant is derivable from bio-
logical laws and the fact that he has been taking birth control pills, but the fact that he has been taking birth control
pills cannot be any part of the explanation for why he
failed to become pregnant.

The restriction to syntactic considerations makes it impossible

1. to establish the explanatory asymmetry between cause and
effect when the relation of constant conjunction is symmetric.
2. to rule out an explanation as inappropriately shallow
because it highlights a mere associated sign, and not a genuine cause.
3. to eliminate useless additions to the premises of the explanation. (Further tampering will not do. As Miller demonstrates in Fact and Method, we cannot devise a principled
way of excluding superfluous elements of explanations
within this model without undermining the superiority of
causally deep explanations in other settings.)

Another method of attack is to point out specific examples of
explanations which are successful, yet fail to involve highly general
covering laws. (Many historical and psychological explanations
fall into this category.)

II.ii From the Inductive-Statistical Model
to the Statistical Relevance Approach

The deductive-nomological model is clearly ill-equipped to deal
with situations where the explanandum is to be expected on the
basis of initial conditions and laws with a degree of probability that
falls short of deductive certainty. Countless explanations in science
are statistical, some irreducibly so (e.g. in quantum mechanics).
Hempel himself acknowledged the need to propose a different, yet
related, model to handle this sort of case: the inductive-statistical
model of explanation. He presumed that a statistical explanation
involves showing that the explanandum is to be expected with high
probability given background conditions and laws.

Hempel added on a requirement of "maximal specificity,"
which states roughly that an explanation must incorporate any rel-

evant information that we could have known before the occurrence of the event to be explained. Though it may not conform ideally to Hempel's model because the statistical regularity appealed to is not appropriately law-like (so much the worse for Hempel's model), the following example illustrates the kind of problem that a condition of maximal specificity is supposed to be of some help in preventing:

Imagine that doctors are discovering an increasingly high rate of infection with HIV among heroin users who share needles. The situation has deteriorated so considerably that nearly 90% of this population is HIV positive. Bob is a heroin user who shares needles. His being infected with HIV is to be expected with high probability given the information we have highlighted. Bob is HIV positive, but it turns out that we have not succeeded in providing the explanation for his infection, because relevant information has been left out. Bob is a hemophiliac who had been taking medication manufactured from HIV contaminated blood, years before he started using heroin. (We shall return to this stubborn problem in our discussion of more sophisticated attempts to describe the formal features of statistical explanation.)[4]

Criticisms of the inductive-statistical model have become as much a part of the common philosophical lore as the criticisms of the deductive-nomological model. Many of the shortcomings of the deductive-nomological model carry over to the inductive-statistical model.

1. We are still unable to rule out an explanation on account of insufficient depth. (A worker at a nuclear power plant developing cancer is to be expected with high probability given background laws and statistical generalizations, plus the fact that a geiger counter he was carrying while investigating an accident site regis-

---

4. Though the requirement of maximal specificity addresses one problem, the need to count Bob's taking contaminated medication as an explanation for his infection, it doesn't in the end help Hempel handle this case with complete success. Bob's being infected is to be expected with high probability on the basis of his being a heroin user who shares needles, and so the latter will—by Hempel's criteria—count as an explanation of the former, even though this verdict is completely at odds with our intuition that it shouldn't.

tered extremely high levels of radiation: several thousand roentgens an hour. In most contexts, the measuring device behaving in a certain way isn't the kind of explanation we desire as we seek to explain why the worker developed cancer.)

2. We cannot eliminate useless premises. (That a fairly substantial earthquake hit LA within the last half of the century was to be expected with high probability given background conditions, statistical generalization and geological theory, plus the fact that Bob has angrily stomped on the San Andreas fault every morning since 1950, but Bob's stomping cannot be a part of the explanation for the earthquake hitting LA in the specified time frame.)

3. We cannot include explanations (e.g. from the social sciences) which do not involve highly general statistical laws. (Consider Thucydides explanation of the Peloponnesian war: Athens generated anxiety by having taken away or threatened the independence of several Greek states. This anxiety, which engendered support for the enemies of Athens, wouldn't have led to war if it weren't for Sparta also feeling endangered.[5]

Furthermore, the specification that the explanandum be expected with high probability on the basis of the explanans generates problems:

1. An explanation in accord with evolutionary theory can be put forward for why a conquering male lion sometimes kills the young fathered by the previous head of a pack, but such behavior is hardly to be expected with high probability on the basis of the observation that killing the young of conquered males ensures that limited resources will be used to promote the genetic material from stronger animals. (After all, there are plenty of species where males never take over and kill the young of their rivals.)

---

5. An advocate of the inductive-statistical model might try to handle this kind of historical case by saying that such an explanation is a mere sketch: we cannot spell out the underlying statistical laws, but the usefulness of the given, incomplete explanation presupposes that such laws exist. This response doesn't seem very satisfying. Are we really to suppose that there are general laws applying to the anxiety of nation states which dictate under what conditions such anxiety leads to war with high probability?

2. Bob develops leukemia a couple of years after being exposed to a radiation level 20 times the annual dose received from natural sources. His exposure to the radiation explains why he developed leukemia, but his developing leukemia was not to be expected with high probability given his level of exposure to radiation. (Among a population of a million people exposed to a radiation level 20 times the annual dose received from natural sources, 30-80 cases of leukemia in the 25 years after exposure are expected.)

Simpson's paradox illustrates a difficulty along these lines as well. If a factor counts as a part of an explanation for a type of event, yet that factor is always or often associated with countervailing forces within a certain population, the presence of the explanatory factor may not raise the probability of the type of event it helps to explain, let alone render the event highly probable. For instance, say that, in a certain community, suffering from a particular kind of childhood trauma has a tendency to cause the development of anti-social behavioral disorders in adulthood, but social workers in the community earmark the children who have suffered from the trauma and enroll them in effective counseling programs. The probability that a trauma victim in the community will develop anti-social behavioral disorders is, on account of the counseling programs, not particularly high (and is in fact equal to the probability of the development of such disorders for a random member of the community's population), yet appeal to childhood trauma is still a relevant part of explaining why anti-social behavioral problems arise in some cases.

The inadequacies of the inductive-statistical model did not end all hopes for developing a formal account of (or at least a foundation for) scientific explanation which would avoid reference to irreducibly causal reasoning. Neo-positivists still anticipated that highly general features characteristic of all scientific explanations could be described, and that, to some extent, formal rules divorced from specifically causal considerations could be laid out which would specify what could count as a genuine explanation and why.

The development of the next and most recent major successor

in the quest for a formal, acausal model of scientific explanation has been heavily influenced by an interest in avoiding the most glaring technical failure of its predecessor. As our last set of examples demonstrated, the requirement that the explanandum be expected with high probability on the basis of the explanans is crude and misguided. In the new proposal, the statistical relevance model of explanation, what matters is not high probability of the explanandum relative to the explanans, but rather, change in probability relative to the explanans. An explanans highlights factors which are statistically relevant to the occurrence of the explanandum, though it doesn't need to render the explanandum highly probable to count as a legitimate part of an explanation.

The statistical relevance model shall be our focus for the remainder of this chapter. Because of its degree of refinement, our paradigmatic example shall be the sophisticated statistical relevance model Salmon presents in *Scientific Explanation and the Causal Structure of the World*(1984), not the original version he advocates in *Statistical Relevance and Statistical Explantion*(1971) even though, in the later publication, Salmon himself rejects his earlier acausalism. (He still insists, however, that causal notions stand in need of analysis– just not a purely statistical one).

In the earlier work, Salmon doesn't devote much space to describing the philosophical motivations which underlie his reductive view, but it is clear enough that he takes for granted the Humean considerations which inspired Hempel. His recognition of the need for a new formal approach stems from his concern with the incapacity of the inductive-statistical model to accommodate certain kinds of cases, and not any deep philosophical misgivings. Similarly, his abandonment of the reductive aspect of the statistical-relevance approach seems largely inspired by the intractability of specific counterexamples, and the inability of purely statistical concepts such as *common cause* or *the screening off relation* to be of much help in these instances.

Though he no longer regards it as the whole story, Salmon would like to think that the statistical relevance model at least serves as a pre-causal foundation for a proper account of scientific explanation. The considerations we shall raise in the next section cast doubt on this more modest claim, as well as the stronger view that the features highlighted by the statistical relevance model constitute the essence of scientific explanation.

III. THE STATISTICAL RELEVANCE MODEL OF EXPLANATION

III.i An Outline of the Model and Some Refinements

In accordance with a somewhat watered-down version of the statistical relevance model, if we are interested in providing all the information which might be useful or needed in a request for a scientific explanation for an event (why a member x of a certain reference class A came to possess a certain property B), we should specify the following[6]:

1. prior probability $p(B/A)$, the probability that an arbitrary member of A will possess B.

2. an explanans partition, a mutually exclusive, exhaustive partition of A $\{D1, D2 \ldots D2^n\}$, where each cell of the partition consists of the members of A that satisfy a particular combination of the properties (or their negation) which are statistically relevant to B.

We form the cells of the partition by considering all and only properties of members of A which are relevant to B $(s1, s2 \ldots sn)$, and forming sets $(d1 \ldots d2^n)$ from these properties by including, for each si, either si or its negation. Taking each cell Di to consist of the members of A satisfying all the properties in di, we end up with our mutually exclusive, exhaustive partition.

3. posterior probability $p(B/Di)$ for all Di.

4. a determination of which cell in $\{D1 \ldots D2^n\}$ x belongs to.

An illustration will make these formal specifications seem more vivid and sensible. Imagine that we wish to provide a com-

---

6. The comprehensiveness of the model is one of its puzzling and questionable features. Note that we needn't regard followers of this approach as insisting that any explanation which fall shorts of providing all this information is ultimately inadequate: the goal is simply to highlight all potentially important information, information which could be valuable or required in some contexts.

prehensive explanation for why a maple tree suffering from a certain disease died from the disease. Suppose, for the sake of example, that the factors that are statistically relevant to this disease
becoming deadly for maples are (s1) exposure to drought, and (s2)
exposure to high levels of pesticides (both of which are positively
relevant). Our reference class A is the set of maples who suffer
from the disease. The event we wish to explain is why a member x
of the class A came to possess property B (death from the disease).
Our reference class partition {D1 . . . D4} consists of:

D1:	the set of all members of A satisfying	s1 & s2
D2:	"	"	s1 &-s2
D3:	"	"	-s1 & s2
D4:	"	"	-s1 &-s2

(We shall assume in this example that the positively relevant factors are so strongly relevant that even in the presence of a negatively relevant factor -si, the net result is positive relevance, i.e.
$p(B/D_i)>p(B/A)$ 1_i_3)

Our rigorous explanation of the event will then include a specification of $p(B/A)$, the probability that a maple suffering form the
disease will die from it, and, for all Di, $p(B/D_i)$ (e.g. $p(B/D2)$, the
probability that a maple that has been exposed to drought, but not
high level of pesticides, will die from the disease). Finally, we must
indicate which partition x is a member of (say in this case it is D2).
Since s1 &-s2 is positively relevant to B, we see that our explanation displays how the possession of property B becomes more
probable by the influence of this combination of factors. In other
words, $p(B/A)<p(B/D2)$. Highlighting the change in probability
does the explanatory work.

The simple outline of a statistical relevance model that we
have looked at captures the most important aspects of the statistical relevance approach to explanation, though a few refinements
and elaborations are in order:

1. In our outline, an explanans partition was included. To be
truly comprehensive, we could also include an explanandum partition. Instead of focusing exclusively on B, the property or event type

that we wish to explain, we set up an exclusive, exhaustive partitioning of attributes that includes that one we are ultimately interested in. For example, in our maple tree example, we could use:

B1: death by disease
B2: complete recovery
B3: partial recovery
B4: no change. Our prior probabilities are then all the $p(Bj/A)$'s, and the posterior probabilities are all the $p(Bj/Di)$'s.

By partitioning the explanandum in this way, we are highlighting the contrast class (death by disease vs. survival or recovery, not death by this disease vs. death by some other disease): we secure which "why question" we're asking. We are also now free to enhance our explanations by noting how certain combinations of factors affect the probabilities of other possible outcomes. (We see that a certain combination of factors leads to an outcome which is highly probable given the presence of these factors, namely death, but how much do these factors allow for partial recovery? for complete recovery?)

2. Salmon recommends that we also specify marginal probabilities $p(Di/A)$. Though it would be too tedious and involved to show why (see Salmon(1984), p.40 for a derivation), these probabilities enable us to ascertain the probabilities of the various Bj's given satisfaction of one or some other combination of factors short of the entirety. We are then able to trace the degree of influence of one of the contributing factors involved. (For instance, in our maple example, we can assess what degree of contribution to the risk of death drought exposure poses.) We can determine to what extent a factor's influence is strengthened or undermined by the presence of other factors.

One could hardly be blamed for wondering what the rationale for incorporating such comprehensiveness in the model in its simplest version and its refinements could be. Salmon's original motivation seems to be the intuition that an explanation of an event simply needs to be this exhaustive if it is to allow us to accomplish what is, for him, the primary task of scientific explanation: to find the place of the event in the pattern of statistical (and, he would now add, causal) relations that make up the history and life of our orderly world. According to Salmon, attaining full scientific understanding of why x is By involves more than focusing narrowly on what properties x has, and how these properties contribute to being or becoming By: we also need to be able to picture how the full range of relevant factors play themselves out towards positively or negatively effecting the occurrence of By (and all other alternatives in the contrast class). We can be satisfied that we fully grasp how this sick maple tree died (exposure to drought, but not exposure to pesticides) only when we know the degree to which different combinations of these conditions contribute to or counteract the effect we are interested in understanding, as well the alternative possible effects in the contrast class.

The demand for this kind of extreme thoroughness marks a further departure from Hempelian ideas. Whether such a departure was ever really necessary is certainly a question which proponents of the model need to address more fully, but not one which we shall dwell on here.

3. We have seen two refinements which have the effect of increasing the amount of information embedded in a bona fide statistical relevance-style explanation. Salmon also adds on a qualification that helps get rid of excess: when two Di's differ only by one conjunct dx, but yield the same probability for Bj, we should merge the relevant Di's into one cell in order to avoid unnecessary divisions in our partition. For example, if the probability for death in sick maples is the same given drought and exposure to pesticides as it is given drought and the absence of exposure to pesticides, we can merge D1 and D2. Absence or presence of exposure to pesticides becomes statistically irrelevant given drought.

4. Just as Hempel ultimately needed to add a condition of maximal specificity in his model of explanation, someone who is trying to construct a statistical relevance model must include a requirement of objective homogeneity: each member of the explanans partition $\{D1..D2^n\}$ must be objectively homogeneous with respect to the Bj's.

We basically included this refinement in our original model

with our instructions that the explanans partition be built up from "all and only properties of members of A which are relevant to B." For full scientific understanding to be achieved, each partition Di must be objectively homogeneous, i.e. there must be no further way of dividing or grouping the members of Di which is relevant to any Bj.

III.ii The Requirement of Objective Homogeneity

A partition Di is objectively homogeneous with respect to attribute Bj if and only if any well-motivated way of creating sub-partitions of Di reveals the same probability of Bj within each sub-partition. Sticking to a frequency interpretation of the probabilities in question may be helpful for grasping this idea. For example, say our reference class A consists of people who died last year in NY city, our partition D1 consists of middle aged males who died last year in NYC, and 'death by heart attack' serves as our attribute B1. The partition D1 counts as objectively homogeneous with respect to B1 iff any appropriate subclass of D1 (such as 'smokers within D1,' 'vegetarians within D1') reveals the same frequency of death by heart attack. We can imagine, in the case at hand, that the frequency or probability of heart attack is greater in the subparti-tion of smokers within D1 than the class of non-smokers within D1, so the partition D1 is not objectively homogeneous with respect to B1. Our explanation for why a man in New York died of a heart attack last year would have to involve more than a ref-erence to his age and sex.

To describe a case of objective homogeneity we could alter our example slightly. Keeping the partition D1 the same as above, let us take as our attribute class B2' 'death by accident involving elec-tric shock.' It is possible to suppose that the frequency of occur-rence of death by accident involving electric shock among all appropriate subpartitions of D2 (the lawyers, the heterosexuals, the fathers . . .) is invariant. (This is not to say that 'death by acci-dent involving electric shock' is never correlated with other fac-tors, just that it isn't for the groups in question.)

Objective homogeneity is important in the development of a

statistical basis for explanation in science because we want to be able to specify what it means to say that all factors relevant to the occurrence of a type of event have been taken into account. If there are factors (creating subdivisions within a partition Di) which effect the probability of Bj, our partitioning does not enable us to know all there is to know about what kinds of events or properties are statistically relevant for Bj. Inhomogeneity amongst the Di's with respect to Bj suggests that the specification of Di is incomplete for Bj as far as explanation is concerned (whether or not the quest for explanation is regarded merely as the search for statistically relevant factors). To remedy this situation, we would need to incorporate the factors which make Di inhomogeneous with respect to Bj into our reference class partition: they are part of a proper and comprehensive explanation of the event in question.

A failure to recognize inhomogeneity could result in our introducing faulty or incomplete explanations, much the same way that failing to heed Hempel's condition of maximal specificity led to error in our HIV case. For example, suppose that, in accordance with recent speculations, high iron levels in the blood do in fact contribute to the development of heart disease, perhaps (for the sake of illustration) far more dramatically than consuming foods high in saturated fat which are not also high in iron. Imagine that the fact that foods high in saturated fats are often also high in iron has prevented us from recognizing the effect of iron until recently. A health practitioner (Howard) who resists paying attention to medical news assumes that among his patients who died (the reference class), the partition D1 of his deceased patients who didn't smoke, exercised regularly, and maintained diets extremely high in saturated fats is homogeneous with respect to the attribute B1, 'died of heart disease.' Guided by his old medical school teachings, he designates the high saturated fat content in the diets of his patients in this group a significant part of the explanation for their developing a fatal heart disease (exercising regularly and not smoking are negatively relevant).

As it turns out in our story, the reference class D1 is not homogeneous with respect to attribute B1. The frequency of heart disease deaths among Howard's patients in D1 whose primary sources of saturated fat were fatty dairy products (such as cheese, which is low in iron) is much lower than the frequency of such deaths among patients whose primary source of saturated fat was meat (which is high in iron). Howard's explanation is wrong: the primary culprit of the heart disease deaths of many patients in D1 is the high consumption of iron, not the high consumption of saturated fat. If Howard subdivided D1 into 'heavy consumers of dairy products but not meat' and 'heavy consumers of meat,' he would have seen that D1 is inhomogeneous with respect to B1 and perhaps resisted his original, incorrect explanation.

Homogeneity of Di with respect to Bj, on the other hand, is an indication of the absence of 'hidden variables' to account for the distribution of Bj within Di.

Talking about objective homogeneity in a general, somewhat sketchy fashion may seem unproblematic, but obstacles do get in our way if we attempt to give a more precise, technical definition. To be true to the motivations for establishing an acausal, statistical relevance model of explanation, we must try to establish what should count as the well motivated or appropriate *selection rules* for subclasses of Dj (within which one searches for invariance of frequency of B) without relying on any irreducibly causal considerations or intuitions. A selection rule is a rule which "picks out" or defines a subclass of Dj: it establishes whether or not any given member of Dj is a member of the subclass. One can then check to see whether the frequency of B in the subclass is the same as it is in the original class Dj.

If we can specify what sorts of selection rules should count as proper and admissible, we are in a position to offer a criterion for when a class Dj is objectively homogeneous with respect to B. Dj is objectively homogeneous with respect to B if the frequency of B ($p(B)$) is invariant amongst all the subclasses of Dj picked out by proper selection rules (the frequency of B in all the subclasses is the same as the frequency of B in Dj).

In a purely formal specification of constraints on legitimate selection rules, however, it is easy to inadvertently rule out selection rules that would show that Dj is inhomogeneous (so Dj might be counted as homogeneous when it's clear it shouldn't be), or, alternatively, to admit selection rules which end up counting Dj as inhomogeneous, even though it's clear Dj should be regarded as homogeneous. As much as Salmon tries to avoid these problems by designing his constraints to handle potential counterexamples, he is not immune to them. I shall try to show how the kinds of imperfections we find in his account reflect serious difficulties any acausal, statistical relevance model must face.

III.iii Salmon's Acausal Criteria for Admissible Selection Rules and Ensuing Problems

To avoid confusion, we should note at the outset that, in his discussion of how to set the formal criteria for admissible selection rules, Salmon counts the members x1, x2, x3 . . . of the relevant reference class partition D as events (in a sequence[7]). This characterization marks a slight, and ultimately insignificant change from the way he talks about reference class members when he outlines the basic features of the statistical relevance model. In both cases, the explanandum is an event, but in the outline of the model the explanadum is simply a member of a reference class acquiring a certain property/undergoing a particular change, and in the specification of admissible selection rules, it is an event having a certain property/falling under a particular sort of description.

Salmon must define which rules for selecting a subsequence from D are proper and relevant to determining the homogeneity of D with respect to B. He must establish what kinds of subsequences need to be checked for invariance of p(B), and what kinds of rules will pick out all the appropriate subsequences.

In order to try to meet these demands, he introduces the notion of an *associated sequence*. An associated sequence Q is a sequence of events y1 y2 y3 . . . which corresponds to the sequence D: x1 x2 x3. . . . One can use the occurrence of C within Q to select a sequence S within D in which one determines the probability of B: if yi is a C, the corresponding xi belongs to S. If, according to at least one admissible selection rule, the

---

7. The emphasis on "in a sequence" is important for handling infinite cases.

selected sequence manifests a higher or lower probability for B than p(B) in D, D is said to be inhomogeneous with respect to B; if all admissible selection rules yield sequences with an invariant probability for B (p(B) in D), D is said to be homogeneous with respect to B.

Salmon then places a couple of constraints on what can count as a legitimate associated sequence, i.e. an associated sequence that can be used to define an admissible selection rule:

1. The y_i must lie in the past light cone of the corresponding x_i: y_i is spatio-temporally local with respect to x_i, and whether y_i is C is independent of x_i or events after x_i. Salmon chooses to express this idea by referring to an ideal detector/computer: "the membership of y_i in C or -C could, in principle, be ascertained by a computer that receives information from a physical detector responding to y_i, but that receives no information gathered by the detector (or from any other source) in response to x_i or to any event z_i in the absolute future of x_i . . ."(Salmon, p. 68) (To avoid confusion later, note that the expression "y_i is spatio-temporally local with respect to x_i" does not mean that the events y_i and x_i are "close" to one another, but rather that they are located in space-time in such a way that a signal could, in principle, be transmitted from one event to another without violating the principles of special relativity.)

2. C must occur within Q in a mathematically random fashion. (For future reference, the following expressions shall be interchangeable: "random associated sequence of C in Q" and "random associated sequence Q consisting of elements y_i which are C or -C." With both expressions I mean to indicate that C occurs amongst the members y_i of Q in a mathematically random way. In accordance with Church's definition of mathematical randomness, I am saying that the probability of C is invariant under all place selections corresponding to effectively calculable functions: no purely ordinally defined gambling system[8] can be devised which

---

8. Purely ordinally defined rules are those which can be applied by taking into account only the positions of the members in the sequence.

would allow you to guess correctly, before time i, that yi will be C with better probability than p(C) in Q.)

We might wonder if it is appropriate to build restrictions against action at a distance into an abstract characterization of statistical relevance, but the scientific motivation behind this condition is clear enough. In general, the justification for the first constraint is fairly straightforward. We need to ban the use of sequences which involve properties of events which are simultaneous with xi, for otherwise we could simply pick out all the B's under some alternative description (for example, we could pick out all the events 'a 2 is rolled' in a sequence of rolls of a fair die by counting C as 'the occurrence of an even prime when the die is rolled'). Such a selection rule would render any class of events inhomogeneous, except for classes which are trivially homogeneous, classes containing all or no B's. We also need to ban the use of selection rules which appeal to properties of future events in counting whether yi is C or -C, for otherwise we would be able to render D inhomogeneous relative to B by admitting selection rules that use effects of B events to pick out B events with higher probability than P(B) in D.

Salmon includes the second constraint in order to stave off the following sort of objection. Imagine a number r expressed as a series of 1's and 0's with a 1 in the ith place for every xi which is a B in D, and 0 in the ith place for every xi which is a -B in D. Suppose a computer is fed the number r (one could think of a lot of computers being fed different ri's, one of which is r). The yi are 'the computer reads the digit in the ith place of r,' and C is 'a 1 is read in this place.' If we use the occurrence of C's in this sequence as a selection rule for picking out events in D, we will have a subsequence of D-events with a much higher probability of B than P(B) in D (namely 1), and so D must count as inhomogeneous with respect to B. This result is undesirable: any sequence corresponding to a computable r automatically must count as inhomogeneous, even if there is no physical basis for partitioning off a class within D in which B is more probable.

Someone may think that this example shouldn't be taken into

account because C and Q don't involve any actual physical events or their properties. We can remedy this situation by using Q to create a new associated sequence Q'. Pick any sequence of events S with elements ji (some of which possess an arbitrary property T, others of which are -T) which lie in the past light cone of the corresponding xi in D. Defining yi and C as above (in the "computer reading r" example), generate an associated sequence Q' (with elements yi' which are C' or -C') in the following way:

yi' is C' iff yi is C and ji is T
   otherwise,
yi' is -C'

Every occurrence of C' in Q' picks out a B, just as C in Q does, but unlike C and Q, C' and Q' do depend on the properties of actual physical events, however strange these properties may seem. The subsequence of D-events picked out by C' in Q' has a much higher probability of B than p(B) in D, so D must count as inhomogeneous with respect to B. Just as in the original example, this result holds whenever B in D corresponds to a computable r.

Salmon eliminates the undesirable consequences of the original "computer reading r" example by specifying that C has to occur within Q in a mathematically random way. For any case where B in D corresponds to a computable r, a selection rule based on C in Q will be banned since C occurs within Q in a mathematically nonrandom fashion (random implies noncomputable). If B in D corresponds to a noncomputable number, we don't have to worry anyway: the idea of a selection rule based on a computer reading a noncomputable number is incoherent.

But couldn't there be a non-random associated sequence which picks out B in D with a particularly high probability, while the probability of B in all subsequences picked out by random associated sequences is invariant? Any non-random sequence is banned from consideration by Salmon's second constraint; hence, under these conditions, D would have to count as homogeneous, even though it ought to be counted as inhomogeneous.

Salmon never considers such an objection, but he needn't

worry—at least not yet. The objection cannot work, for given any nonrandom associated sequence of C in Q which picks out B in D with probability PDQ(B) (the probability of B in the subsequence of D picked out by the associated sequence Q), it is always possible to construct a random associated sequence of C' in Q' which picks out B with at least as high a probability as PDQ(B). Here is how the construction should proceed if we presuppose that we have a nonrandom associated sequence Q consisting of elements y_i which are C or -C. Take any random associated sequence S consisting of elements j_i which are T or -T, where all the j_i lie in the past light cone of the corresponding x_i. Construct an associated sequence Q' (with elements y_i' which are C' or -C') in the following way:

yi' is C' **iff** yi is C and ji is T
    otherwise
yi' is -C'.

C' in Q' will pick out B in D with at least as high a probability as PDQ(B). (The subsequence of D picked out by Q' is contained in the subsequence of D picked out by Q, so PDQ'(B)≥PDQ(B).) Since the generation of C' in Q' is dependent upon the elements of a random sequence, C' in Q' is itself noncomputable and random.

This strategy may spare Salmon from one kind of problem, but it also gives rise to a new difficulty for his statistical relevance model. I have introduced a technique for sequence construction which I have used twice: first, to show how the "computer reading r" example (which justifies the inclusion of a randomness constraint in the first place) can be modified to include associated sequences consisting of actual physical events, and second, to respond to a potential criticism of Salmon's randomness constraint. The very same technique, however, can be used to show that the randomness constraint is ultimately useless: it is powerless to eliminate the kind of problem it was designed to solve.

We have just seen how we can generate a random associated sequence which picks out B with at least as high a probability as a the probability of B in the subsequence picked out by a given nonrandom associated sequence. By making a substitution, we can use these same instructions to slip past Salmon's randomness constraint and renew the threat posed by a modified "computer reading r"-type example.

Let "yi," "C" and "Q" stand for the same thing they stood for in the modified "computer reading r" example (C in Q corresponds to the 1's in the binary representation of B in D). "T," "S" and "ji" stay the same (S is a random associated sequence with members ji which are T or -T).

Our instructions show us how to build a random associated sequence Q' with members yi' which are C' or -C':

yi' is C' **iff** yi is C and ji is T
   otherwise,
yi' is -C'

The following illustration may help readers visualize how this construction proceeds:

	1	0	0	1	1	
D:	B	-B	-B	B	B	...
Q:	C	-C	-C	C	C	...
S:	T	T	T	-T	T	...
Q':	C'	-C'	-C'	-C'	C'	...

   ▲   ▲

PDQ'(B) = PDQ(B) = 1, but, unlike C in Q, C' occurs in Q' in a random way, and hence the associated sequence of C' in Q' cannot be ruled out by the randomness constraint. Assuming D is not trivially homogeneous with respect to B, whenever B in D corresponds to a computable r, D must be considered inhomogeneous with respect to B. P(B) is not invariant under all selections by legitimate associated sequences, for it is always possible to construct a random associated sequence of C' in Q' such that PDQ'(B)=1.

A "computer reading r" selection rule belongs to the broader class of ordinal place selection rules (instructions for picking out subsequences which make use of the position of elements in a series). Salmon thinks that the randomness constraint can successfully eliminate such selection rules from homogeneity evaluations, but we have seen that he is wrong.

To be fair to him, we must acknowledge that he sometimes wonders whether such a constraint is always necessary, particularly for finite cases where it could have no application. Recall we are using Church's definition of mathematical randomness in sequences, invariance under all selections which correspond to effectively calculable functions– a notion which applies only to infinite sequences. The constraint can't work for finite sequences, but aren't these the kinds of cases we usually care about?

For finite cases, Salmon suggests, perhaps it is enough to replace the randomness constraint with the specification that an admissible associated sequence must be purely physically determined: the members y_i of an associated sequence C must be physical events, and whether or not y_i is C or -C must be a purely physical fact. Ordinal place selections are banned from consideration. This exclusion is particularly appropriate for finite cases where the order of a sequence doesn't have to be important (order is extremely important for infinite cases: without a specific ordering the idea of an infinite sequence is incoherent).

This suggestion does nothing to avert a simple, but serious problem that seems to face any acausal statistical relevance model which is supposed to cover finite cases, the cases we should be most interested in having our model accommodate. If our only constraints on what can count as an admissible associated sequence Q are: 1. that the members y_i lie in the past light cone of the corresponding x_i; 2. that the members of Q be physical events, and the property C which is used to define a selection rule be a physical property of members of Q, we can be sure that *any* finite sequence will turn out to be inhomogeneous with respect to *any* property of its members. Let "D" stand for a finite sequence D, and "B" stand for a property that some of the members of D possess. Consider the binary expansion which corresponds to B in D (a 1 stands in the ith place for every x_i which is B, a 0 stands in the ith place for every x_i which is -B). Given the astronomically high number of physical events and sequences of events that lie in the past light cone of the x_i, there are bound to be many event sequences that have exactly the same binary expansion. Any one of these associated sequences Q_j could be used to pick out B in D perfectly: PDQj(B)=1. Correspondingly, any one of these associated sequences could be seen as highlighting past events or event types which are statistically relevant to the occurrence of B in D. But who would want to invoke these past events in an explanation of B in D when all that may tie the sequences together is a shared binary expansion?

I really cannot see a way of solving this problem without making use of causal notions. Adding time constraints– by specifying, for example, that the yi can't have occurred in the distant past of the xi or in a place far away from the xi–is not an effective solution. Sometimes events which are causally (and statistically) relevant to other events did happen long before and far away from the events which we wish to explain. As far as I can tell, the only way to prevent statistical relevance from collapsing into an empty notion–a notion which is useless for purposes of explanation–is to invoke the idea that an associated sequence can only be admissible if its members yi lie in the causal nexus of events leading up to the xi. Of course, such a solution is no solution for anyone who hoped to use statistical relevance as a foundation for clarifying what "cause" really amounts to, or dreamed of finally purging explanation of its apparent dependence on flighty causal notions.

III.iv Difficulties in the Identification of Causal Relevance With Statistical Relevance

The purest version of the statistical relevance model of explanation forces us to identify causal relevance with statistical relevance, yet there are plenty of types of examples (above and beyond the one we have just explored) which seem to show that such an identification is misguided. Statistically relevant factors can't always be thought of as causally relevant, and causally relevant factors can't always be though of as statistically relevant. Here are some illustrations:

A. Statistically relevant factors which are not necessarily causally relevant, but are included by the model

Salmon shows how a classic example that would seem to fall

under this category isn't, in the final analysis, a threat to the statistical relevance approach to causality and explanation.

s1: The atmospheric pressure drops sharply.

s2: The barometric reading drips sharply.

B: A storm occurs.

$p(B/s2) > p(B)$

At first it appears that s2, which is statistically relevant to B, would have to count as a proper cause or at least part of an explanation of B within the statistical relevance model, even though many of us have strong intuitions that s2 isn't really a cause of B; it's just a sign of a proper cause. Recall, however, that Salmon includes a special requirement for partition merging: we cannot make divisions in the explanans partition which have no bearing on the probability of the explanandum when the partitions resulting from such a division differ by only one conjunct. The factor s2 can be "screened off" by the factor s1 [$p(B/s1 \& s2)=p(B/s1)$]; hence, s2 cannot be included in a genuine explanation of B, nor counted as a proper cause of B.

We see how partitions formed from conjuncts s1 and s2 would have to be merged, so that s2 is eliminated from consideration:

$p(B/s1 \& s2) = p(B/s1 \& -s2) = p(B/s1)$
$p(B/-s1 \& s2) = p(B/-s1 \& -s2) = p(B/-s1)$

The two initial partitions in each row differ by only one conjunct, where this conjunct has no bearing on the probability of the explanandum B.

The factor s2 cannot even potentially screen off s1 because $p(B/s1 \& s2) \neq p(B/s2)$: barometers sometimes fail to work properly.

The screening off trick has worked this time, but it cannot always help. There are several types of situations where a causally irrelevant (or "less strongly relevant") factor or event is statistically relevant to the explanandum, but the model can't reject it or deem it inferior.

Example 1

Say a factor s2, which is causally irrelevant to B (and statistically relevant to B only in virtue of s1, i.e. s2 is irrelevant to B in the absence of s1), supervenes on s1 (which is causally and statistically relevant to B).

> s2 supervenes on s1: s1 → s2
> s2 is statistically irrelevant to B in the absence of s1:
> p(B/-s1 & s2) = p(B/-s1)

Imagine, for example, that a gene sequence which always leads to purple eyes is responsible for the development of a certain disease. Purple eyes, which can also result from different gene sequences, themselves do not play any role in the development of the disease.

> s1: The agent has the gene sequence.
> s2: The agent has purple eyes.
> B: The agent suffers from the disease.

Since we stipulated that s2 is statistically irrelevant to B in the absence of s1, there is no question that -s1 screens of s2, and partitions {-s1,s2} and {-s1,-s2} can be merged. In cases where we state the cause of the development of the disease in the absence of the gene sequence, purple eyes wouldn't be invoked.

The same cannot be said for cases where the disease develops in the presence of the gene sequence. In a sense, s1 screens off s2 (p(B/s1 & s2) = p(B/s1)), but our model will not let us eliminate s2 as a cause or part of the complete explanation for B, even though purple eyes have no causal bearing on B. The partition {s1, -s2} is empty (since s1 implies s2); consequently, p(B/s1 & -s2) is undefined. We cannot then say p(B/s1 & -s2) = p(B/s1 & s2) = p(B/s1), and merge {s1, -s2} with {s1, s2}. When s1 is present, s2 cannot be eliminated–a very unattractive and counterintuitive consequence of following the rules of the statistical relevance model.

<u>Example 2</u>

If the relation between s1 and s2 were stronger than supervenience, namely if s2 were a perfect sign of s1 (s1 iff s2, i.e. s2 when and only when s1), no merging of partitions in virtue of conjunct s2 would be possible since partitions {s1, -s2} and {-s1, s2} would both be empty.

Say purple eyes were always perfectly correlated with the gene sequence which is responsible for the development of the disease mentioned in example 1. If we adhere to the model strictly, purple eyes would always have to count as a proper cause and part of a complete explanation for the development of the disease.

Other kinds of cases follow roughly the same pattern illustrated in examples 1 and 2. Consider instances of events that have a causally shallower (s2) and causally deeper (s1) explanation. Sometimes the relation between the shallower and deeper causes is one of supervenience: the shallower cause supervenes on the deeper cause, as in the case of mental causes (the mental state supervenes on the neurophysical state). Here, example 1 is relevant, with the qualification that s2 need not be considered statistically irrelevant to B given -s1 (s2 cannot be screened off, even with respect to -s1).

When s2 is reducible to s1 (the way facts about heat are reducible to facts about molecular kinetic motion), the pattern outlined in example 2 is followed. There is no way, within the model, to isolate out the deeper cause as deeper: both shallower and deeper explanations are on a par as far as the model is concerned, and reference to both is a necessary part of a complete explanation (however redundant the shallower cause may seem in some contexts[9]).

Of course an advocate of the power of statistical relevance could insist, at this juncture, that—at least with respect to cases where a shallower and deeper cause is involved—the model simply provides all potentially important information about the causes of an event. Context dependent, pragmatic considerations dictate which part of this information is genuinely useful or appropriate.

---

9. For example, in legal contexts in which the M'Naghten rules are followed, the claim that someone is insane is reducible to the claim that the person is ignorant of the moral quality and character of his acts, without being in a position to try to correct for or avoid this ignorance. In an explanation of why the accused is considered not responsible for his actions, and should be spared the usual sort of punishment for the crimes involved, the claim about ignorance is arguably more significant and revealing than the claim about insanity.

I'll grant that this response goes some way towards answering worries about the model's neutrality with respect to shallower and deeper causes, though I think that the results of the original examples 1 and 2 remain highly counterintuitive. The same holds for example 1 below, which could be rightfully be placed under category A and category B.

B. *Causally relevant factors which are eliminated by the model*
Example 1
Say s1 can be causally responsible for B, and $p(B/s1) > p(B)$, but s2 is an event which is highly correlated with, but prior to B, so $p(B/s2)$ is almost 1, and $p(B/-s2)$ is almost 0.

> $p(B/s2 \ \& \ s1) = p(B/s2 \ \& \ -s1) = p(B/s2)$
> $p(B/-s2 \ \& \ s1) = p(B/-s2 \ \& \ -s1) = p(B/-s2)$
> s1: Daryl Strawberry hits a bad pitch using his "special trick."
> s2: The struck baseball is high in the air, heading towards the bleachers, 1 inch from the outfield boundary and inside the foul line.
> B: A homerun hit is accomplished.

Unfortunately for fans of statistical relevance, s2 screens off s1, so s1 cannot be considered a relevant cause or part of a complete explanation, even though it should be. In addition, s2 is highlighted as a cause or part of a complete explanation for B, even if s2 is not causally responsible for B under any conventional understanding of what "causal responsibility" amounts to.

Example 2
A prior cause s1 can be screened off by a later cause s2 when s2 is a highly effective means for bringing about B given s1 (and would be just as effective in the absence of s1 as well).

> $p(B/s1 \ \& \ s2) = p(B/-s1 \ \& \ s2) = p(B/s2)$
> s1: Bob is extremely angry at his computer monitor and wants to destroy it.
> s2: Bob runs over the monitor with his pickup truck.
> B: The monitor is demolished.

The statistical relevance model is prone to incongruent tendencies: sometimes it is too permissive, leading too much to what supporters would call "pragmatics" and retractors would call "reasoning about causes"; at other times, it is too strict, leaving too little. The fit between statistical relevance and causal relevance is at best imperfect, and at worst fairly irrelevant to a deeper understanding of the foundations of causal intuition.

The model may go some ways towards illuminating how we should make judgments about what counts as a good explanation, but its failings seem to confirm the contemporary explanationists' suspicion that the rules we follow when we make considered choices on such matters depend on irreducibly causal principles, yet tend to resist general description. In the end, realists have nothing to fear from the statistical relevance approach and its fallen predecessors.

CHAPTER THREE

Van Fraassen's Arguments Against Inference to the Best Explanation

Bas Van Fraassen, one of the most interesting and innovative empiricists of our time, has been a harsh critic of inference to the best explanation, presenting a host of arguments against the idea that the explanatory superiority of a theory can be a reason for regarding the theory as rationally compelling, or more likely to be true than explanatorily inferior, but equally empirically adequate rivals. After a brief overview of his own version of anti-realism, and why it may be found wanting, we shall examine these arguments (including his most elaborate and threatening challenge to explanationism to date, the Bayesian Peter objection), and see that their shortcomings render them incapable of undermining the most reasonable forms of explanationism.

I. VAN FRAASSEN'S CONSTRUCTIVE EMPIRICISM

Van Fraassen has moved well beyond the linguistic prejudices of his positivist predecessors, rejecting the idea that causal talk (or, to use Earman's expression, "weasel causal words") must be translated into phenomenological equivalents or statements of statistical relevance. He proposes, instead, a "semantic view of theories." According to this view, a scientific theory is to be identified with a family of models, a set of structures which satisfy the basic claims or postulates of the theory. Two empirically equivalent rivals may say the same thing about observables—the portions of their models which map onto the observable part of the world may all be isomorphic—but they can still present contrasting pictures of what the unobservable realm is like.

Van Fraassen also disavows the positivist view that confirmation, or deciding which among a range of contrasting causal accounts is best, is simply a matter of conforming to highly general, a priori canons of good reasoning. Though all scientific inference is subservient to the ultimate aim of securing empirical adequacy, the specific interests and circumstances of the inquirers play a major role in determining a favored direction for research. Pragmatic considerations settle which among contrasting, equally empirically adequate theories should be pursued, and which aspects of available accounts of the causal nexus leading to the event to be explained are salient.

Faithful to the credo of traditional empiricism (information we receive through our senses serves as the sole foundation for rational inquiry), van Fraassen regards empirical adequacy and consistency as the only features of theories which are relevant to our judgments of likelihood of truth. In his view, we have no reason to regard any given theory about unobservables as more likely to be true than any one of its equally empirically adequate rivals–rivals which always exist in principle. A particular theory may be chosen over these rivals because it provides the best explanation for the data at hand, or serves as the wisest option given the competing pressures influencing scientific research, but such a choice is always governed by purely pragmatic considerations, irrelevant to likelihood of truth. We are, then, never rationally compelled to believe any theory about unobservables, even one which the scientific community universally adopts; at best, we are constrained only to accept it as empirically adequate. This modest requirement accords well with van Fraassen's conception of the paramount end which drives scientific inquiry: the pursuit, not of a greater understanding of the workings of nature, but of useful, empirically adequate theories.

Van Fraassen feels that the demand for empirical adequacy, logical consistency, and a further requirement for structural compatibility in one's evolving belief state called "probabilistic coherence" (which I shall explain later when we get to the Bayesian Peter objection) alone comprise the constraints of rationality. Explanatoriness never enhances likelihood of truth, nor does it force us to count a particular hypothesis as a uniquely rational or non-dogmatic belief. This is not to say, however, that actually believing the best explanation is irrational. In accordance with van Fraassen's vision of rationality as a concept of permission rather than compulsion, belief in the best explanation is merely ill-advised.

In an effort to dispel concerns that his form of empiricism runs the risk of being incapable of warding off skeptical despair on the one extreme, and relativist promiscuity on the other, van Fraassen expresses his sympathy with a voluntarist or pragmatist attitude towards epistemology. The skeptic claims that any beliefs going beyond the immediate content of our experience are irrational because such beliefs lack definitive justification: we really cannot demonstrate that our ordinary view of what gives rise to our current sensations is more likely to be true than radical, alternative causal accounts (Descartes' demon, Berkeley's God, Putnam's mad scientist...). The pragmatist, however, thinks that this stringent requirement on rational belief is unnecessary. Any philosophical theory about the proper exercise of reason and epistemic caution rests on some choices and assumptions about what the ends and virtues of inquiry are, or what sort of cognitive conduct best secures these ends, but who is to say that the skeptics' stance on these issues must have sway over us? We may decide to value courage and imagination over never being wrong, so adopting hypotheses without being able to rule out all possible alternatives, once and for all, need not always strike us as unreasonable.

Correspondingly, the pragmatist can try to resist the impression that "everything is permitted" by noting that epistemic and non-epistemic projects alike are subject to evaluation in a number of different ways which depend on what we perceive as goodness or success in the given enterprise. Just as a sculptor's efforts to create a bust of Shakespeare may be criticized for being too conventional, too peculiar, too hasty, too cheap with materials... so can we find reasons for regarding a particular theoretical construction as an instance of shoddy or unappealing workmanship. It is well within our right, for instance, to look upon the alien abduction hypothesis of Harvard psychiatrist John Mack with disdain because we do not admire the recklessness of being so willing to embark on speculative flights of science-fiction fancy when the psychological well-being of troubled people is at stake, and more conventional explanations for the phenomena in question are readily available.

Many of us, however, will find this approach to making room for the criticism of theories counterintuitive. We would like to count the acceptance of a set of beliefs as epistemically irrational and irresponsible precisely when we detect that an agent's interest in trying to secure the truth is clouded over by deep psychological needs and desires. For the pragmatist, however, there is no sense in which following norms which reflect a good strategy for cooperatively striving towards the truth is more "in order" or epistemically responsible than following norms which reflect a good strategy for securing a feeling of well-being, omnipotence, enhanced self-esteem and superiority to others... The only epistemic rule is that empirical adequacy and consistency are to be respected—a rather minimal constraint given the fact that, with enough imaginative tinkering and ad hoc maneuvering, virtually any view can be molded to save the phenomena. If feeling superior to other people is genuinely of upmost importance to a person (and who is in a position to deny that, for some sorry souls, success in this project may be the key to happiness and contentment?), there is nothing within pragmatism itself which produces the verdict that this individual is violating the rules of good reason if she allows all her choices in inquiry to be directed towards securing what she values the most.

Van Fraassen leaves room for theoretical criticism, but he counts only those critiques which highlight problems with a theory's logical consistency or empirical adequacy as truly epistemic and impartial. In general, a criticism of a theory should only have weight amongst people who share the critics's conception of what sorts of values, interests and goals should properly inform research and theory construction. This view, as we have noted, is counterintuitive in its permissiveness. Products of dogmatism, wishful thinking or rampant speculation can all count as rationally acceptable and beyond reproach from an impartial, epistemic standpoint so long as they are internally consistent and strain to fit the observable facts.

In opposition to this unappealing stance, realists have tried to show that a host of considerations beyond consistency and empirically adequacy—considerations relevant to the assessment of explanatory strength and superiority—can play a role in a fair, impartial assessment of whether belief in the theory is irrational, rationally acceptable or rationally required. The most promising efforts in this direction, as I have emphasized earlier, fall into two categories:

(i) Externalist, naturalist explanationism. Purely a posteriori methods akin to those in the natural sciences are used to defend realism, and a reliabilist account of why explanatoriness is epistemically relevant is offered: judgments of explanatory superiority are informed by background theoretical considerations which are, in fact, largely true.

(ii) Internalist, non-naturalist explanationism. Additional constraints on rational belief are revealed by reflecting on the preconditions for the possibility of meaningful inquiry into a subject matter and the cooperative pursuit of understanding. Such preconditions include the acceptance of basic evidential principles without which grasp of relevant concepts would be impossible. Explanatoriness can count as an epistemic virtue when the greater explanatoriness of a theory is grounded in the theory's superior conformity to these truistic commitments.

Van Fraassen has issued a counterattack, and it is up to us to decide whether the explanationists' efforts to avoid the dire consequences of van Fraassen's pragmatist, constructive empiricism have any hope of surviving his arguments against inference to the best explanation.

II. BUILDING A CASE AGAINST EXPLANATIONISM: THE SHORT ARGUMENTS

II.i The Scientific Image

Although *The Scientific Image* is often taken as van Fraassen's classic statement and defense of constructive empiricism, the arguments against explanationism which it contains are surprisingly brief and limited in scope. His most extensive critique of realist uses of inference to the explanation, while nicely complementing his discussion of the pragmatics of explanation, is aimed specifically at a somewhat obscure feature of early forms of explanationism which is not present in important contemporary views.

Van Fraassen begins his tirade by highlighting problems with J. Smart's cosmic coincidence argument for realism. Smart claims that we must take highly empirically successful theories about unobservables to be true, or else their capacity to accommodate relevant past, present, and future observables would have to be seen as an amazing cosmic coincidence. The phenomenological rival T' saves the phenomena as well as T, but belief in it cannot be a rational substitute for belief in T: T' taken by itself leaves too many regularities unexplained. Believing in the truth of T allows us to explain regularities that stand in need of explanation, including the empirical success of T. Believing T' alone, however, prevents us from being able to account for the continued empirical success of T (not to mention the observed regularities which are explicitly within the scope of T and T').

As van Fraassen notes, it cannot be the case that belief in T' alone is inadequate simply because T' leaves some regularities or other unexplained. After all, any theory we can think of at the present time, no matter how causally rich, will leave some regularities or other unexplained. We may succeed at accounting for the behavior of unobservable entities at a remarkably distant level of causal depth (think of the progression from macroscopic phenomena to molecules to atoms, to electrons, protons and neutrons to subatomic particles to superstrings...). But unless a great innovator in theoretical physics can come up with a convincing, remarkable reason for thinking otherwise, it looks like we will always have to settle for accepting that the most causally deep entity in our available repertoire behaves in a certain manner for which we have no currently acceptable explanation.

By failing to describe or acknowledge any sort of defensible limit to the demand for explanation, Smart seems to leave us with two unattractive options. We can regard all our scientific theories as forever unfinished and unsatisfying, the products of a Sisyphean striving towards infinity, or we can admit that the empiricist stance is inevitable, in which case we may as well adopt it right from the start and concede that belief in T' alone is fine.

Van Fraassen favors the latter alternative, though he accepts that a realist wouldn't have to cave in to the empiricists at this juncture if she had a good way of preventing the demand for explanation from becoming a categorical imperative. He considers one significant attempt to provide a formal rule for determining what sorts of regularities genuinely stand in need of explanation, Reichenbach's principle of the common cause, but this approach turns out to have special problems of its own. It leads to the positing of hidden variables in quantum mechanics, a result which is incompatible with current theory and the results of certain important experiments.[1]

Van Fraassen has put forward a legitimate challenge to any defender of realism: if you insist that some theoretical explanations must be believed on pain of irrationality, let alone that some phenomena intrinsically stand in need of explanation, you have to be prepared to give an account of how to set reasonable, nonarbitrary limits to the demand for explanation. Such an account may depart from the utter generality we see in Reichenbach's proposal and

---

1. According to Reichenbach's principle of the common cause, when coincidences occur that are too improbable to be attributable to chance, they must in principle be explainable (and should be explained) by appeal to a common cause. For example, environmental scientists are quite concerned about the population reduction and increased incidence of deformity amongst frog species throughout the world. The problem seems too widespread and severe to assume that the assaults on the various species are entirely unrelated. At least some of these species must be subjected to similar kinds of destructive elements in their environment (a common cause for the severe declines in population and increases in defects).

Formally speaking: when two events A and B occur more frequently together than we would expect given their independent likelihood of occurring, i.e. when A and B satisfy the relation $p(A\&B) > p(A)p(B)$, then there exists a common cause C which gives rise to A and B, and satisfies the following conditions:

(1) C precedes A and B

(2) $p(A/C) > p(A/-C)$ and $p(B/C) > p(B/-C)$

(3) $p(A\&B/C) = p(A/C)p(B/C)$

A B and C form what is called a "conjunctive fork."

allow for reference to topic or circumstance-specific considerations so long any appeal to pragmatic or context-dependent features does not undermine the special, epistemically privileged status of the theories in question.

I think that this challenge can be met successfully by both the externalist and internalist explanationist. The externalist can claim that, as with the evaluation of explanatory adequacy and superiority, background theoretical considerations play a significant role in the determination of what needs to be explained in the first place. Theoretical heritage dictates, for instance, that the law of inertia should be taken as a foundational principle for the investigation of motion. To accept that we need to use the law to explain other phenomena rather than explain the law itself is to be faithful to the part of classical mechanics which still has value (and, the externalist would want to add, is approximately true).

The internalist explanationist would put a slightly different gloss on this example. We do make use of specific causal principles embedded in current theory when we make choices about what stands in need of further explanation and what does not, but what makes these choices non-arbitrary and appropriate is not necessarily the approximate truth of background theory, but rather the very content of the concepts involved. The relevant notions of motion, force and velocity on their own sustain the requirement that the law of inertia be taken as explanatorily fundamental.

To grasp a concept of an object, we must be able to place the object in the relevant scheme of reasons, evidence, explanations, inference rules, and so on. A person who understands a concept must have a tendency to reason with it in the ways which are most fundamental to its content. This capacity consists of more than being able to determine what serves as evidence for facts about the objects which fall under the concept, or what provides good or bad reason for forming beliefs about the objects. It also involves knowing what sorts of facts concerning the object stand in need of explanation, and what kinds of explanations of these facts could count as a fully satisfying, thus ending a chain of demands for deeper and deeper causal explanations which might otherwise extend off into infinity. Thinking that we can, in the proper circumstances, adequately account for someone's anger by pointing out that she was insulted, or adequately explain why someone is sad by noting that her good friend had recently died is part of understanding the concepts of anger and sadness. To believe that causally deeper explanations are always in principle required is to violate the constraints which issue from the content of the concepts themselves.

After discussing the need for realists to set reasonable limits to the demand for explanation, van Fraassen turns to his next target in *The Scientific Image*, arguments for realism which involve appeal to the general success of science (not simply the success of any given empirically adequate theory as in Smart's case for realism). Boyd's abductive argument for realism is the most sophisticated and well-developed version of this argument, an approach to defending realism with serious flaws (see chapter V), though not necessarily any which are successfully identified in the *Scientific Image*.

In response to the claim that only realism can adequately account for the amazing empirical success of science (or, as Boyd would wish to emphasize, the amazing instrumental success of the theory-dependent methods of science), van Fraassen proposes that the anti-realist can meet the explanatory challenge just as well. If we look at nature, we see that it's no surprise that the species of the world have features which are well suited for survival in their habitats: the species which don't have such features become extinct. Similarly, it is no surprise that science and its theory-dependent methods are empirically successful, since theories which fail to be empirically adequate get weeded out.

At first blush, van Fraassen's "Darwinian," anti-realist explanation for the phenomenon in question seems to be just as acceptable and appropriate as a Darwinian explanation for the explanandum highlighted in the argument from design. We don't need to posit the existence of a Great Designer in order to explain how living things have such intricate features which allow them to thrive in their environments: species which fail to develop such features fail to survive. Similarly, we don't need to see our scientific theories and theory-dependent methods as truth-tracking in order to explain how they could have features which make them empirically successful. Theories and methods which fail to have these features don't survive.

Upon further reflection, however, we can start to wonder whether this explanation addresses the appropriate explanandum. The sort of success Boyd feels we need to account for is perhaps not empirical success *per se*, but empirical success with a certain character, or under a certain description—a contribution to instrumental reliability which is, according to Boyd, so astonishingly pervasive and effective that we have to think that some special causal mechanism must be at work. His explanatory challenge could not then be regarded as analogous to the one raised by the argument from design, but rather, to a demand for explanation which might appear in evolutionary biology when thriving is unexpected or puzzling. Consider the question "why are marsupials capable of flourishing in Australia, when they have failed to compete successfully with the better adapted placental mammals in the rest of the world?" In this case, it is obvious that the nonspecific, Darwinian response which was appropriate to answering the theistic challenge ("the marsupials have the sorts of features which enable them to flourish in their environments; otherwise, they would all be dead") will be completely unsatisfying and uninformative—more a rejection of the need for explanation than a response to it. If the phenomenon Boyd wants us to explain really is more like the explanandum in the marsupial case than the explanandum in the argument for design, we can criticize van Fraassen's anti-realist explanans on the same grounds and insist that it too is completely unsatisfying and uninformative.

Van Fraassen could, of course, change tactics at this juncture and argue that Boyd's explanandum is ill-conceived: there's nothing to explain because the theory-dependent methods of science haven't actually been so astonishingly effective. Such a strategy is not one he chooses to pursue, however.

II.ii. Laws and Symmetry

Though the most menacing anti-explanationist argument in *Laws and Symmetry* is presented in the parable of Bayesian Peter, van Fraassen makes some introductory points in order to show what sorts of general considerations have influenced his rejection of inference to the best explanation. He begins by expressing his dissatisfaction with an implicit belief on the part of those philosophers who regard inference to the best explanation as rationally compelling: "For me to take it that the best of set X will be more likely to be true than not, requires a prior belief that the truth is already more likely to be found in X, than not." (Van Fraassen 1989, p.143) He goes on to evaluate what he regards as the three possible ways of defending or modifying this assumption within a realist framework, and finds that each of these options is unsatisfying.

1. *Privilege*: The realist might try to take what van Fraassen calls a "naturalist" stance and argue that we have evolved to be predisposed to select an appropriate range of hypotheses, a range within which the truth is more likely to be found than not. This approach is highly dubious. Survival seems to require at best only that we have a capacity to develop empirically adequate theories about the most commonplace aspects of the world. There is no reason to suppose that this skill has any bearing on an ability to approach formulating true (not just empirically adequate) theories about deep, complex features of the universe.

2. *Force majeure*: A realist could also attempt to avert the issue of whether we have reason for thinking that the truth is more likely to be found amongst our range of available alternatives than not, and insist that the act of choosing an alternative by itself entails believing it. In other words, independent of whether or not we have good reason to think that the truth is likely to be found in our given set of hypotheses, and independent of the fact that we have no choice but to make a unique selection from the available options if we wish to move forward, adopting a preference for a particular explanation and sticking with it is basically the same as believing it. Our actions show that we all follow a rule of inference to the best explanation.

This response is no more promising than the previous one: being forced to make do and work with what's available—however epistemically impoverished—bears no doxastic obligation or content. Just because we in fact choose to leap over a wide crevice to avoid being eaten by a bear doesn't show or imply that we believe that we will survive the dangerous jump.

3. *Retrenchment*: A realist might also try to argue that inference to the best explanation has been misconstrued. The real rule does not dictate that we infer to the truth of the best explanation, but rather that we take the degree of explanatoriness into account when we think about the degree of likelihood of a hypothesis. According to this view, we are rationally constrained to regard good explanations as possessing special features which, either before or after the evidence comes in, make them more likely to be true.

Van Fraassen defers attacking the idea that we are rationally required to take explanatoriness into account when we update our personal probability assignments in light of new evidence (he feels his Bayesian Peter objection is needed to take care of this sort of explanationism), though he does make a few remarks to undermine the general view that special features of good explanations, considered independently of what future evidence might arise, make superior explanations more likely to be true. He starts by making what some realists may regard as a somewhat startling assumption about common ground: "I believe, and so do you, that there are many theories, perhaps never yet formulated but in accordance with all evidence so far, which explain at least as well as the best we have now." (van Fraassen 1989, p.146) These theories are in large part incompatible with one another, so most of them must be false. We know nothing concerning the truth value of our best explanation beyond the fact that it belongs to this class of theories, most of which are false. Our explanation is just a random member, so we must regard it as very improbable that it is true.

In his discussion of 1-3 van Fraassen makes some reasonable points. We can certainly agree that we have no reason to think that we are by nature bound to select an appropriate range of hypotheses at every advanced stage of inquiry, nor that choosing to work with what seems like the most attractive hypothesis amongst what might be fairly meager options is itself an act of believing it. We can also be sympathetic with van Fraassen's skepticism about the idea that there are shared features amongst all best explanations which make them likely to be true–after all, if there were such features, why have so many of our best explanations turned out to be false? In some cases, we are well aware that our investigative resources are extremely limited, leaving us with the active possibility that there are a great number of attractive alternatives to the best available choice which we haven't been able to develop or consider yet. Our best option is simply the best among a poor selection, so we have no reason for thinking that it is likely to be true.[2]

Despite these legitimate observations, the impact of van Fraassen's critique is undermined by his failure to consider more promising realist responses to the charge that accepting the rational force of inference to the best explanation involves defending the indefensible claim that we have reason to believe that the truth is more likely to be found in our range of available alternatives than not. Some of these responses issue from realist projects which, like Boyd's, may have serious shortcomings, but these failings are independent of any of the difficulties van Fraassen raise here.

The externalist explanationist appears to adopt what van Fraassen calls the "privilege" option. She appeals to naturalism in order to secure epistemic privilege for our range of available alternatives, though she doesn't need to rely on the evolutionary story which van Fraassen sees as a fundamentally flawed part of the naturalist approach. For the externalist, judgments of explanatory superiority in science are guided by (if not equivalent to) judgments of theoretical plausibility: plausibility in light of what she regards as approximately true background theory. Our best explanations in science, and the sets of potential alternatives from which these best explanations are drawn, tend towards approximate truth because we arrive at them by means of causal considerations

---

2. For a fuller discussion of why we shouldn't think that, in general, the best explanations is likely to be true, see chapter VI.

which issue from approximately true background theory. She doesn't need to think that evolution has made any significant contribution to our capacity to attain this sort of foundation for future inquiry (perhaps scientists of old were just lucky when they first started to land on the right track). She can, like Boyd, argue that belief in such a foundation is a part of the externalist, naturalist package which can best account for the sorts of phenomena a philosophy of science ought to accommodate (including the instrumental success of the theory-dependent methods of science).

The externalist explanationist would also take issue with van Fraassen' assumption about common ground in the discussion of the retrenchment option. She would not accept that there are always a number of scientific explanations, inaccessible to us now, which could in principle serve as genuine, satisfactory alternatives to the ones we currently use. Causal considerations from approximately true background theory help us narrow our range of possibilities to the most reasonable, likely options amongst the many equally empirically adequate alternatives.

The internalist explanationist has the means for rejecting van Fraassen's common ground assumption as well. An explanation sometimes counts as best because it conforms uniquely well to truistic, criterial evidential and causal principles—principles which must be accepted by anyone who is to count as genuinely possessing the relevant concepts (in this case, concepts which are a part of our most basic efforts to cognitively grasp the world). Belief in the principles (and, consequently, belief in explanations which they endorse) is rationally required, for being able to grasp the relevant, basic concepts is a prerequisite for being able to participate in cooperative inquiry into the world. There are not always, in principle, unformulated alternatives to these truistically endorsed explanations which would provide equally adequate explanations to the current ones: we already see that the current explanations follow from the application of our truisms to experience, and we have no reason to think that some currently unavailable alternatives could mesh as well with these criterial principles.

The internalist explanationist is not committed to the general validity of inference to the best explanation. She simply holds that inference to the (approximate) truth of the best explanation can be rationally compelling in some cases, namely those cases where explanatory superiority is grounded in truistic, criterial considerations which issue from the ordinary concepts we use to cognitively grasp the world in the most basic way. This version of realism could accommodate an even more moderate stance toward inference to the best explanation, however—a stance towards which the objections van Fraassen has raised up to this point do not apply. It would still be perfectly within the realist camp to say that truistic considerations rationally compel us to regard the best explanation not necessarily as more likely to be true than false, but rather, more likely to be true than incompatible, available alternatives—even when such alternatives are equally empirically adequate. (By adopting this more moderate view, we don't have to worry about whether unformulated, unavailable alternatives would or would not be able to mesh with our truisms as well.) Explanatoriness, when it issues from basic criterial principles, is epistemically relevant, and should have a bearing on our probability assignments, over and above what the raw empirical data dictates.

The externalist version of realism, it should be noted, can also be recast in this light. Degree of explanatoriness should at the very least have a bearing on our assessment of comparative likelihood of truth, beyond what the raw evidence dictates, because explanatoriness is, in Boyd's words, "evidential." Judgments of explanatory superiority are guided by largely true theoretical principles drawn from approximately true background theory, so an explanation which explains well (i.e. is plausible in light of background theory) should be accorded a higher probability than any conflicting, equally empirically adequate alternative which does not.

Since the objections van Fraassen has raised up to this point have no bearing on these moderate forms of explanationism, and pose, at best, a limited threat to stronger forms (undermining only the most extreme view—that we are always rationally constrained to believe the best explanation—as well some patently implausible ways of trying to defend it), explanationism has fared quite well so far. Can it survive the next round?

III. THE BAYESIAN PETER OBJECTION

III.i Van Fraassen's Dutch Book Argument

By targeting the moderate form of explanationism (explana-toriness should have a bearing on our probability assignments, over and above what the raw empirical data dictates), the second major objection in *Laws and Symmetry* poses a potential threat to most anyone who feels that inference to the best explanation plays a central role in confirmation. In this rather technical polemic which is, strangely enough, presented in parable form, van Fraassen argues that an agent who allows explanatoriness to enhance her degree of credence is susceptible to a Dutch book. She will have to regard as individually fair a series of bets which lead to a certain loss. This state of affairs indicates that accepting explanationism leads to an unacceptable incoherence in one's belief state.

Developing a Dutch book is a way of showing how certain epistemic decisions are irrational or incoherent—in violation of the axioms of probability. An implicit assumption in such an approach is that any time a bookie could use an awareness of your principles for probability assignment to sell you a series of individually fair bets which would in combination represent a sure loss for you, regardless of outcomes, your assignments (and the principles dictating such assignments) are irrational. Given that a fair price for a bet on A which pays W if you win (0 if you don't) is $p(A)W$, a bookie can determine which bets you would consider fair on the basis of your degrees of belief and rules for degree of belief revision. If he can then offer you a series of individually fair bets which would lead to a certain loss, your belief state as a whole must contain a structural inconsistency.

The character who represents the explanationist victim in van Fraassen's Bayesian Peter parable is hapless Peter who, influenced by a preacher of inference to the best explanation who tells him that the explanatory power of a hypothesis allows him to raise his credence, makes use of an ampliative rule according a weight to a hypothesis greater than that dictated by coherence as evidence starts to accumulate. Peter is confronted with an alien die with a bias of any one of $X(1)$, $X(2)...X(10)$, he knows not which (a bias of $X(n)$ indicates that the probability of ace on any single proper toss is $n/10$). His opinions about the likelihood of various outcomes—such as the die coming up aces for four consecutive tosses—are at this point in strict conformity to what would be suggested by the probability calculus (his personal probability for the die coming up aces four times in a row would then be the average of the probability of this outcome for each of the ten possible scenarios, bias $X(1)$, bias $X(2)$, etc.). A friend comes along, and they negotiate bets which are individually fair, in accordance with Peter's probability assignments. In the course of their game, a string of four aces is tossed. Thinking that the best explanation of this occurrence is a hypothesis of high bias, Peter raises his current probability for the event "the fifth toss shows ace" turning out true from .87 (the result obtained from a routine application of Bayes' Theorem) to .9. After further betting negotiations with his friend (which again are fair according to Peter's updated probability assignments), he winds up with a Dutch book.

This overview of the Bayesian Peter Objection is extremely sketchy and condensed, but I think it contains all you really need to know to understand the last two of the three criticisms I will raise. Since my first criticism will target some of the technical details of the argument, here is a more thorough version of the story:

As we mentioned above, Bayesian Peter is confronted with an alien die with a bias of any one of $X(1)$, $X(2)...X(10)$, he knows not which (again, a bias of $X(n)$ indicates that the probability of ace on any single proper toss is $n/10$). A friend comes along, and proposes some bets:

> Given proposition E: the first four tosses of the alien die show ace and proposition H: the fifth toss shows ace
> Bet I pays $10,000 if E is true and H is false
> Bet II pays $1300 if E is false

Bet III pays $300 if E is true.

At the beginning of the die tossing, when Peter has the same initial probabilities as his friend, they both evaluate the fair cost of each bet in accordance with the laws of the probability calculus.

> The initial probability that E is true = the average of (.1)4, (.2)4,... ,(1)4 = .25333 (Van Fraassen's ending with (.9)4 in the text is just a typographical error: the initial probability for E which he states is correct.)
> Initial probability that E is false = 1-.25333= .74667
> Initial probability that E is true and H is false = the average of (.1)4(.9),...,(.9)4(.1), 0 =.032505
> The fair cost of a bet that pays x if y is xp(y) so:
> Bet I costs $325.05
> Bet II costs $970.67
> Bet III costs $76.00

Peter buys all three bets from his friend ($1371.72). We see that if all four tosses do not come up ace, Peter loses bets I and III, and wins bet II. In this case, Peter's losses amount to 1,371.71 - 1,300= 71.72.

Van Fraassen asks us to assume that E turns out true. Peter wins III and loses II, so he is paid $300. So far he has a net loss of $1,071.72. At this point, information from an earlier part of the story which I haven't included yet becomes relevant. Before running into his friend, Bayesian Peter was confronted by an IBE preacher who convinced him that the explanatory power of a hypothesis allows him to raise his credence. Peter came to believe that posterior probability can be fixed by prior probability, outcomes, and explanatory success.[3]

In light of this belief and the string of four ace tosses, which is best explained by a hypothesis of high bias, Peter raises his probability for H being true from .87 (the result obtained from a routine application of Bayes' Theorem[4]) to .9. Peter sells bet I to his friend for $1,000 (the payoff is $10,000, and the probability that H is false is .1).

---

3. See *Laws and Symmetry*, p. 166.
4. See *Laws and Symmetry*, p. 169.

Thanks to the influence of the IBE preacher, Peter winds up with a Dutch book. Now that he has sold bet I, he must count his net losses at 1,071.72 - 1,000 = 71.72. He could have seen at the beginning of the game that he was bound to lose 71.72, whether or not E and H turned out to be true or false.

Van Fraassen triumphantly warns: "What is the moral of this story? Certainly, in part that we shouldn't listen to anyone who preaches a probabilistic version of Inference to the Best Explanation, whatever the details. Any such rule, once adopted as a rule, makes us incoherent." (Van Fraassen 1989, p. 169)

We don't even need a complicated story about the follies of Bayesian Peter, however, to see that it might be difficult to conform to the probability calculus as evidence comes along once we allow explanatory power to influence our judgments about likelihoods or likelihood comparisons. Take a look at Bayes' Theorem, an important rule for updating personal probability assignments in light of new evidence (p is your old probability function, p' you new one after evidence e comes in, and H1...Hn is an exhaustive list of the possible alternative hypotheses):[5]

$$p'(Hi) = p(Hi/e) = \frac{p(e/Hi)p(Hi)}{p(e/H1)p(H1)+p(e/H2)p(H2)...+p(e/Hn)p(Hn)}$$

Since the denominator is the same for all the p(Hi/e)'s, we can compare the p(Hi/e)'s just by looking at the p(e/Hi)p(Hi)'s. Imagine a situation where H and H' are both accorded the same initial probability (say, for example, that prior to exposure to startling

---

5. Here's a quick, informal derivation of the theorem so you have some idea of where it comes from: Bayes' theorem regulates passage from your old probability function p, to your new probability function p' in light of evidence by the principle of Bayesian conditionalization. The principle can be applied when we commit ourselves to new evidence e (p'(e)=1).

p'(h) = p(h/e)

(e gives support to h when p(h/e) > p(h)).

We start out with p(a & b) = p(a/b)p(b) = p(b/a)p(a)

(A rough way to see that this makes sense: when two events are independent, i.e. p(c/d)=p(c) and p(d/c)= p(d), their joint probability is just the product of each taken individually. In a sense the a & b event is equivalent to (a/b)b or (b/a)b.)

We want to determine p(hi/e). So far we have:

evidence, both are considered equally likely), e is considered equally likely on H and H'—that is, p(e/H)=p(e/H') (e.g. e is a deductive consequence of both hypotheses given certain background assumptions)— and H is a better explanation for e than H'. We might be tempted to say p(H/e)>p(H'/e) because of H's capacity to explain e in a more satisfactory way, but a quick glance back at Bayes' theorem shows us we've run into trouble: all the values on the right side of the equals sign would be the same, while the values on the left side would be different.

Someone might try to avoid this difficulty by protesting that anything that would go into the evaluation of H as a better explanation than H' would already be reflected in a higher prior probability for H than H'. We could imagine, for example, that strange spaceship-like traces were found on the golf course of an exclusive, high security retirement community. The residents form two hypotheses:

> H: outer-space creatures have stopped by for a visit
> H': a retiree has pulled an elaborate prank.

Given the genteel air of the community, the members decide that they are all too serious and reserved to have gone to great lengths to make it look as though a spaceship has landed in the golf course, so they select H as the best explanation for the puz-

---

$$\text{*}\quad p(h_i/e) = \frac{p(h_i \,\&\, e)}{p(e)} = \frac{p(e/h_i)p(h_i)}{p(e)}$$

To get p(e), the probability of e before it happens, we use the theorem of total probability (whenever we have a list of incompatible, exhaustive alternatives $h_1...h_n$, i.e. $p(h_1 \vee h_2 \vee ...h_n)=1$, any other probability statement can be expressed as a weighted sum of likelihoods). Applying the theorem to the case at hand, we begin with an exhaustive list of incompatible, alternative hypotheses to be subjected to test, $h_1...h_n$ (where $\forall i \; \forall j \; i \neq j, h_i \neq h_j$).

$$p(e)= p(e/h_1)P(h_1) + p(e/h_2)P(h_2)...p(e/h_n)P(h_n)= \Sigma p(e/h_i)p(h_i)$$

Substituting in * above, we obtain

$$p(h_i/e) = \frac{p(e/h_i)p(h_i)}{\Sigma p(e/h_i)p(h_i)}$$

zling state of affairs. p(e/H)= p(e/H'), but p(H/e) being greater than p(H'/e) can be accounted for without violating Bayes' theorem by noting that, in this example, p(H) > p(H'). Our protestor would say that explanatory superiority here is grounded entirely in greater initial probability.

This suggestion works very well for this example, and may prove to be quite helpful as we try to square explanationism with Bayesianism; it appears to have some limitations, however. The fact that (under the interpretation of the function we have been using) p(H/e) > p(H'/e) even though p(e/H) = p(e/H') cannot always be grounded in differences in prior probability assignments to H and H.' Imagine that we regard e1 and e2 as incompatible, unusual, but possible events which are equally likely. e1 and e2 are equally likely on rival theories H and H' which in turn are accorded the same probability, but H explains e1 better than H' explains e1 and H' explains e2 better than H explains e2 (H and H' focus on different types of phenomena). We don't know what evidence we'll end up facing, so which hypothesis is most explanatory can't be settled ahead of time and reflected in prior probability assignments p(H) and p(H'). Here we haven't a chance of tracing preferences based on explanatory power to prior probability assignments since p(H)=p(H'). In this example, p(H/e1) > p(H'/e1), but p(e1/H)=p(e1/H') and p(H)=p(H').

How, then, are we to respond to van Fraassen's Bayesian Peter objection, and the presumably related, but more general charge that explanationism is incompatible with Bayes' Theorem?

III.ii Flaws in the Argument

As I see it, there are three major difficulties with van Fraassen's argument.

1. *Peter's Carelessness*: Van Fraassen nonchalantly states at the beginning of the tale that, before any evidence comes in, Peter will calculate his personal probability for E&-H, the object of Bet I, by taking the average of p(E&-H/bias X(1)),...,p(E&-H/bias X(10)). Using $p_n(Y)$ as shorthand for p(Y/bias X(n)), and applying the axiom P(A&B)=p(A)p(A/B), we see that p_n(E&-H)= p_n(E)p_n(-H/E).

Under each assumption of bias $X(n)$, the probability that the first four consecutive tosses comes up ace, but the fifth does not is $(n/10)^4(1-n/10)$. Van Fraassen then calculates Peter's value for $p(E\&-H)$ in the following way.

$p(E\&-H) =$
the average of $p_1(E\&-H),...,p^{10}(E\&-H) =$
the average of $p_1(E)p_1(-H/E),...,p^{10}(E)p^{10}(-H/E) = \dfrac{\Sigma(i/10)^4(1-i/10)}{10}$
the average of $(.1)^4\,(.9),...,(.9)^4(.1), 0 = .032505$

Because $p(E\&-H)=p(E)p(-H/E)$, Peter's values for $p(E\&-H)$ and $p(-H/E)$ are interdependent. Given $p(E)$ and one of these values, the other is determined. Using the values for $p(E)$ and $p(E\&-H)$ above, $p(-H/E)= .032505/.25333 = .13$. But we know, from what transpires later on in the story, that $p(-H/E) = .1$. For Peter, once E happens, the hypothesis of high bias gets "bonus points" on account of its explanatory success, and this increase in credence in turn affects his personal probability for the fifth toss being, or not being, an ace.

Van Fraassen's calculation of $p(E\&-H)$ on Peter's behalf takes no accounting of the fact that Peter is the sort of explanationist whose probability values for fifth toss events like H or -H will be affected by whether or not E happens. As such, the calculation is not really true to the rules which govern Peter's degree of belief assignments. Given the necessary interdependence between $p(E\&-H)$ and $p(-H/E)$, any determination of $p(E\&-H)$ will somehow have to incorporate Peter's explanationism, a requirement which is violated by van Fraassen's method, the averaging of values for $p_n(E)p_n(-H/E)$ given equally weighed biases.

Who knows exactly what peculiar methods or rules Peter *would* want to use to set a value for $p(E\&-H)$, but we do know from what happens towards the end of the tale that $p(-H/E)=.1$. Consequently, since we know $p(E)=.25333$, we can establish $p(E\&-H) = p(E)p(-H/E)=(.25333)(.1)=.025333$. Even at the start of the game, Peter's fair price for Bet I, which pays $10,000 if E&-H, will then be $253.30 (not $325.05, as van Fraassen assumes). If we use the correct value, and follow along with the rest of Peter's adventures, he will end up gaining 3 cents! (Before he embarks on his final transaction and sells Bet I for $1,000, he has spent a total of only $999.97, not $1,071.72 as in the original version of the parable.) Three cents is a fairly insubstantial amount, but it is enough to save him from the perils of Dutch books.

In the discussion before the Bayesian Peter story is presented, van Fraassen creates the misleading impression that his argument will show that any "ampliative" rule of inference to the best explanation leads to incoherence: allowing explanatoriness (and not just predictive capacity) to factor in as a feature relevant to revising our degree of belief as evidence comes along will, in general, be irrational. The Peter of the original parable is hardly an adequate representative of this sort of broad explanationist view. Instead, he turns out to be a particularly unlikely sort of philosophical victim whose real problem is not his explanationism per se (his belief in the relevance of explanatoriness to degree of likelihood), but rather his tendency not to think very carefully ahead of time of what his rules for degree of belief revision actually are. If the Peter in van Fraassen's version of the story had been attentive to his own rules for updating his probability function in light of evidence, his explanationism wouldn't have been ignored in his determination of p(E&-H), and he would never have been duped in the first place.

2. *An Implausible Explanationist Rule*: We should hardly expect that a real explanationist would, or would have to be, as foolhardy as Peter. Ignoring your own rules for probability function revision at the beginning of betting even though they are relevant to assessing subjective probability values for particular events which form the object of bets, and then letting these rules influence your probability assignments and choice of further betting arrangements later on (in a predictable way) clearly creates opportunities for clever bookies. But what of the explanationist rule which Peter first disregards, and then follows? Do real explanationists even accept this rule?

The rule in question is highly general: the explanatoriness of a hypothesis should make you increase your degree of credence, even beyond what would be recommended by what could be called a "neutral" application of the probability calculus. When you update your probability function in an alien die case, for instance, where the competing hypotheses *high bias*, *low bias* start out with the same prior probability, you should not simply take into account how each hypothesis weighs the evidence, i.e. how probable the evidence is under the assumption that the hypothesis is true; you should add on bonus points once a hypothesis becomes explanatory. This increase in credence will in turn lead to additional increases or decreases in your probability assignments for events (such as, in the case at hand, *the fifth toss shows ace*) which are statistically affected by whether or not the hypothesis is true.

The credence boost in this instance seems highly unmotivated. When we actually make a judgment about which hypothesis, *high bias* or *low bias*, is the best explanation in light of the evidence, there is no information relevant to our decision beyond how each hypothesis weighs the evidence. All that grounds the explanatory superiority of the *high bias* is its greater predictive success, so we would hardly expect to be able to raise our credence over and above what is suggested by a routine, neutral application of Bayes' theorem.

The most reasonable explanationists will recognize that there are many cases, especially in normal science, where hypotheses—all with the same prior probabilities—are essentially distinguishable by the different weights which they accord to various sorts of outcomes. In such cases, the best explanation will simply be the hypothesis which achieves the greatest predictive success. Special bonus points for being the best explanation will not be added (though the best explanation will still be taken to be more likely to be true than the alternatives on normal Bayesian grounds). The reasonable explanationist, then, would not agree that we should follow the rule in the Bayesian Peter story.

This is not to say that the explanationist would never recommend a boost in credence beyond what would be suggested if we simply took predictive success into account. Explanatory superiority is a property sometimes wholly grounded in greater predictive success, sometimes not. When it is not (and there are different views about what more it could consist of: greater plausibility in light of background theory for the externalist, greater fit with truistic, criterial considerations for the internalist), we might find ourselves with additional information which is relevant to our probability assignments. That the explanationist would feel that such a situation can arise is particularly clear when we look at the sorts of examples which most notoriously divide realists and anti-realists: cases where we are faced with incompatible, but equally empirically adequate theories with different degrees of explanatoriness. The realist/explanationist, against the anti-realist/anti-explanationist, will insist that there are cases where we are sometimes rationally constrained to regard the best explanation as more likely to be true than the available alternatives, even when these rivals are equally empirically successful.

In the next section, we shall explore whether there is any incompatibility between this view and Bayes' Theorem.

3. *The Illegitimacy of Diachronic Dutch Book Arguments*: Even some hard core Bayesians, such as Howson and Urbach, believe that only synchronic Dutch Books—Dutch books built up from simultaneous degree of belief ascriptions, and bets the agent simultaneously regards as individually fair—highlight genuine failures of rationality. 'Coherence,' according to this view, is a term which has legitimate application just when used to describe belief states at particular points in time, as opposed to sets of belief states over stretches of time. It is no more incoherent to make degree of belief assignments at different times which correspond to bets which ultimately cannot be regarded as simultaneously fair than it is inconsistent to believe p at one point, and -p later. The probability calculus, then, like the laws of logic, places rationality or coherence constraints on the structure of our belief state at a given moment; both formal systems have much less to say on how our belief state at one time should relate to our belief state at a later time to preserve rationality.

On the assumption that this view is correct (an assumption I defend in the next chapter), we have yet another reason for dismissing van Fraassen's Bayesian Peter objection, which takes the form of a diachronic Dutch Book argument.

III.iii Reconciling Explanationism with Bayes' Theorem

We have noted that both the externalist and internalist explanationist accepts that there are cases where greater explanatoriness warrants a boost in credence above and beyond what we would assign if we took only predictive success or empirical adequacy into account: we are sometimes rationally constrained to regard the best explanation as more likely to be true than equally empirically adequate rivals. We are rationally constrained, for example, to think that the best explanation for the phenomenon of Brownian motion ($e1$)—namely, the molecular hypothesis ($H1$)—is more likely to be true than the hypothesis that matter is continuous ($H2$), even if this radical hypothesis could in principle be incorporated into a non-discreet theory of matter which is empirically equivalent to molecular theory. In other words, once we witness and accept $e1$ ($p'(e1)=1$), our revised probability function p' must assign a higher value to $H1$ than $H2$: $p'(H1)=p(H1/e1) > p'(H2)=p(H2/e1)$, even though $p(e1/H1) = p(e1/H2)$. If we accept Bayes' Theorem, how is this possible?

We have already explored one suggestion which will usually work quite well: accommodate for the differences in value between $p(H1/e1)$ and $p(H2/e1)$, despite the equality of $p(e1/H1)$ and $p(e1/H2)$, by making the $p(H1)$ higher than $p(H2)$. For the externalist this will mean that the prior probabilities $p(H1)$ and $p(H2)$ will reflect H1 and H2's differing degrees of plausibility in light of background theory (independent of considerations of fit with data); for the internalist $p(H1)$ and $p(H2)$ will reflect H1 and H2's differing degrees of fit with truistic, criterial principles (again, independent of considerations of fit with data).[6]

---

6. In the discussion which follows I deal with the externalist and internalist's accounts of what "more, of epistemic significance, there is to good explanation " simultaneously, since the differences in their views has no bearing on any of the points I make. All that matters here is that both regard explanatoriness as sometimes relevant to degree of credence, even when the explanatory superiority of a favored hypothesis is grounded in more than its predictive success. The reader should keep in mind, however, that the externalist and internalist explanationist have contrasting conceptions of what can make explanatoriness epistemically relevant: for the externalist, it is conformity to background theory; for internalist, it is conformity to truistic, criterial principles. Neither one implies the other.

We might come across a situation, however, where which hypothesis is ultimately most plausible in light of background theory or fits the best with truistic considerations will depend on what sort of empirical data comes along. Say that two hypothesis H1 and H2 are, before the evidence comes in, equally plausible in light of background theory, or equally recommended by criterial principles: $p(H1)=p(H2)$. Two possible, incompatible outcomes of a particular experiment are e1 and e2, and $p(e1/H1)=p(e1/H2)$, $p(e2/H1)=p(e2/H2)$, but H1 will fit better with background theory/truisms than H2 if e1 happens, and H2 will fit better with background theory/truisms than H1 if e2 happens (H1 will emerge as the best explanation if e1 happens; H2 will count as the best explanation if e2 happens). Since better fit with background theory/truisms (greater explanatoriness) should have a positive impact on our degree of credence, we want to say that $p'(H1) > p'(H2)$ if e1 happens, or $p'(H2) > p'(H1)$ if e2 happens, but how can we square this probability assignment with Bayes' Theorem, given that all the relevant prior probabilities and likelihoods are the same? Must we abandon our explanationist inclinations in this example and concede that $p'(H1)=p'(H2)$?

For the externalist and internalist, whether H1 or H2 turns out to be explanatorily superior—i.e. is most plausible relative to background theory, or conforms most successfully with truistic, criterial principles—is as relevant to our degree of belief revisions as whether e1 or e2 happens. It would be quite natural for them, then, to acknowledge that the information about which hypothesis possess the highest degree of explanatoriness (fits the best with background theory/truisms) is as significant a constraint on our evolving probability function as the raw evidence itself. We will need to take this new constraint into account if we are to have any hope of successfully applying Bayes Theorem to the case at hand.

Just as we specify a range of significant outcomes for the raw evidence, e1 and e2, we specify a range of outcomes for the new constraint "information about explanatory superiority":

d1: H1 is the best explanation—it fits the best with background theory/truisms once the evidence comes in d2: H2 is the best explanation—it fits the best with background theory/truisms once the evidence comes in.

We will then conditionalize on one of e1&d1, e1&d2, e2&d1, and e2&d2 (depending on what happens) when we update our probability function from p to p'. So, for example, if e1&d1 turns out to be true, $p'(e1\&d1)=1$, $p'(H1)=p(H1/e1\&d1)$, and $p'(H2)=p(H2/e1\&d1)$. We can use Bayes' theorem to calculate these latter two values:

$$p'(H1) = p(H1/e1\&d1) = \frac{p(e1\&d1/H1)p(H1)}{p(e1\&d1)}$$

$$p'(H2) = p(H2/e1\&d1) = \frac{p(e1\&d1/H2)p(H2)}{p(e1\&d1)}$$

Now we see that there is no longer a difficulty with maintaining explanationist inclinations and respecting Bayes' Theorem for the problematic sort of case: $p'(H1) > p''(H2)$, since $p(H1)=p(H2)$, and $p(e1\&d1/H1) > p(e1\&d1/H2)$ (explanationists will think that it is more likely that e1 happens and H1 turns out to be the best explanation if H1 is in fact true, than if H2 is in fact true).

Neither van Fraassen's flawed Bayesian Peter argument, nor more general probabilistic considerations show that commitment to explanationism results in incoherence—the failure to abide by the rules of reasoning suggested by the probability calculus.

We have now examined the major arguments against the explanationist approach to realism which van Fraassen has contributed to the debate, and found that contemporary explanationism at its best has survived them all. Van Fraassen's critique of inference to the best explanation does not successfully demonstrate that explanationism leads to probabilistic incoherence or other equally dire philosophical problems. Instead, the most significant moral to be drawn from these arguments is that explanationists need to fill in the details of their position by establishing what creates demands for explanation and what sets limits to these demands, by placing appropriate constraints on inference to the best explanation, and by specifying when explanatoriness allows you to boost your degree of belief and how this boost should be incorporated into our changing belief state.

CHAPTER FOUR

Van Fraassen's Dutch Books

In the previous chapter I noted that, although van Fraassen's Bayesian Peter objection to explanationism is dismissible on wholly independent grounds, one possible way of undermining it is to cast doubt upon the legitimacy of diachronic Dutch book arguments in general. Here I would like to pursue this line of criticism further and show that, although we may regard vulnerability to synchronic Dutch books as an indication of an epistemically problematic incoherence in one's belief state, failure to sustain "diachronic coherence" does not necessarily reveal any failures of rationality. Diachronic Dutch book arguments are incapable of highlighting new constraints of rationality, or demonstrating that certain rules for degree of belief revision are epistemically troubling and irrational.

I shall argue that the considerations which lend support to the view that synchronic coherence is rationally required do not have any bearing on the issue of whether diachronic coherence is rationally required. Since these considerations do not carry over to the diachronic case, and alternative, general explanations for the requirement of diachronic coherence have not been proposed, it becomes an open question whether rational agents are in fact bound by constraints of diachronic consistency. At the very least, then, we should expect that cases of diachronic incoherence accord with our pre-theoretic intuitions about what constitutes a failure of rationality. As we shall see, however, a number of alleged constraints on rational degree of belief assignment and revision which are derivable from diachronic Dutch Book arguments (van Fraassen's principle of reflection, a related principle which I shall call the "temporally extended principle of reflection," and a prohibition against assigning a probability value to certain kinds of conditional propositions) are particularly counterintuitive and unmotivated. In some instances, violating these principles seems to be precisely what reason demands. The apparent reasonableness of the occasional break from diachronic coherence creates a strong prima facie case for thinking that diachronic Dutch Books arguments do not have the epistemological import their proponents claim they have.

We begin with a discussion of what a Dutch book argument is, and why we might be rationally constrained to comply with the dictates of the probability calculus.

I. A PHILOSOPHICAL APPLICATION OF THE PROBABILITY CALCULUS: USING DUTCH BOOK ARGUMENTS TO DERIVE RATIONALITY CONSTRAINTS

The mathematical investigation of probability has proven to be extremely fruitful when applied to natural science. As the study, in part, of the laws governing the nature of large groups of random phenomena (i.e. events that may or may not take place, relative to a specification of circumstances), it gives us tools for solving problems concerning statistics and stochastic processes raised by quite a broad spectrum of human interests: insurance, ballistics, demography, quality control, economics, measurement error, the kinetic theory of gasses, quantum mechanics...

So flexible is the concept of probability, so varied are its uses that any attempt to demand a single appropriate understanding of what we really mean when we refer to this relation between a random event and its relevant specified circumstances would appear to be forced. We see this sort of strain in the work of von Mises, for example, who insisted on defining probability narrowly as limiting frequency, the limit of the frequency of an event as the number of times its circumstances are repeated approaches infinity. He dismissed both the classical notion of probability ("equiprobability," built up from considerations of symmetry), and, in typical positivist fashion, the need to inquire into what underlying structures of phenomena explain the possession of particular objective probabilities.

Though the occasional neo-positivist might still find favor with von Mises' anti-realist sympathies, most philosophers today would count at least some notion of subjective probability among the range of legitimate interpretations. It is this definition, probability as degree of belief or degree of credence, which has seemed to find a home in philosophy, particularly philosophy of science. The benefit of importing the probability calculus into philosophy of science, as opposed to science itself, is a subject of current debate, but interesting potential applications continue to emerge.

One such application is the use of Dutch Books (sets of individually fair bets which, when taken in sum, lead to certain loss) to point out rationality constraints. Dutch book considerations have been put forward to achieve this end in two different ways:

1. by showing that, much as systems of logic tell us what kinds of abstract relations must hold between our beliefs in order for us to be consistent, the probability calculus dictates rules that our degrees of credence must conform to in order for us to be coherent. To count as rational agents, we have to obey the axioms of probability[1] (p is any personal probability function):

  i. For any proposition A (for which p is defined), $p(A) \geq 0$
  ii. If A is a logical truth, $p(A) = 1$
  iii. If A and B are mutually exclusive, $P(A \lor B) = p(A) + p(B)$

Included as a definition of conditional probability:

  iv. If $p(B) \neq 0$, then $p(A/B) = \dfrac{p(A \& B)}{p(B)}$

By the Dutch Book theorem, if our degree of belief function fails to conform to the axioms of probability, a Dutch Book can be devised against us.[2]

2. by demonstrating that particular kinds of probability assignments or rules for degree of belief revision lead to incoherence. Given that a fair price for a bet on A which pays W if you

---

1. For probability functions defined over an infinite domain it is necessary to add the principles of continuity and additivity. We shall restrict ourselves to considering only discrete cases here, however, since they are much simpler.

2. For proofs of the Dutch Book theorem, see Earman p.39 or Howson and Urbach p.79.

win (0 if you don't) is p(A)W, a bookie could determine which bets you would consider fair on the basis of your degrees of belief and rules for degree of belief revision. If he could then offer you a series of individually fair bets which would lead to a certain loss, your belief state as a whole must contain a structural inconsistency.

Objections to the use of Dutch Books in determining violations of rationality have been raised. A particularly familiar criticism is that although only individually fair bets are offered in the creation of Dutch Books, rational agents can take many different stances towards risk. Even if your personal probability for A is p(A), that doesn't mean that you are always willing to buy a bet on A with a payoff of W for p(A)W, nor that it would be irrational for you to reject such an offer; you might be risk averse. If you resist buying the bets that are offered, why should you care that their sum leads to a certain loss? As long as your degree of belief doesn't reflect a willingness to purchase bets which are determined fair on the basis of this probability assignment, the objection goes, nothing has been said so far about why you can't violate the probability axioms and still be rational.

Howson and Urbach introduce what seems to me an adequate response to this problem for certain kinds of Dutch Book arguments (such as the sort referred to in 1 above).[3] In putting forward a Dutch Book argument, we should not depend upon drawing a dubious connection between degree of belief and action by maintaining that a belief in A to degree p(A) implies that one would buy a bet on A with payoff W for p(A)W. A belief in A to degree p(A) does imply, however, that one would at least regard a bet on A costing p(A)W with payoff W as fair, i.e. as offering zero advantage to either side of the bet, given the way the world appears to the agent:

> agent's side: net gain of W(1-p(A)) if A
>   net loss of p(A)W if -A
> bookie's side: net gain of p(A)W if -A
>   net loss of W(1-p(A)) if A).

---

3. See Howson and Urbach, p.79.

If the agent simultaneously regards particular individual bets as fair (offering neither advantage nor disadvantage), the sum of such bets should also offer neither advantage nor disadvantage. In the case of Dutch Books, however, it turns out that the sum, by guaranteeing a net loss if purchased, offers a definite disadvantage and hence cannot be fair. If a Dutch Book has been developed from bets the agent simultaneously regards as individually fair, we see that there must have been a clash in her original simultaneous degree of belief assignments, a lack of coherence amongst those values which dictate which bets she counts as fair.

In light of the Dutch Book theorem, this line of thought provides an excellent justification for the view that our simultaneous degree of belief assignments have to conform to the axioms of probability in order to be rational. This is not to say that all uses of Dutch Book arguments in accordance with 2 above have been vindicated, however. A Howson and Urbach-style defense of the view that the probability calculus is a source of rational constraint works well for securing the epistemic import of synchronic Dutch Books, Dutch Books which are built up from simultaneous degree of belief ascriptions and thereby bets the agent simultaneously regards as individually fair. Diachronic Dutch Book strategies, on the other hand, are (as the name suggests) based upon knowledge about general rules for degree of belief change over time deliberately followed or rejected by the agent, so Howson and Urbach's defense no longer applies.

According to Howson and Urbach, it is no more irrational and problematically incoherent (in general) to make degree of belief assignments at different times which correspond to bets which ultimately cannot be regarded as simultaneously fair than it is inconsistent to believe p at one time, and then later believe -p. The probability calculus, like the laws of logic, places rationality constraints on the structure of our belief state at a point in time, but both mathematical systems have much less to say on how our belief state at one time should relate to our belief state at a later time to preserve rationality.

This limitation holds even in the case of standard Bayesian

conditionalization, despite Lewis and Teller's well known diachronic Dutch Book argument that, if you are following an updating rule, failing to change your degree of belief $pt0(A)$ to $pt1(A)=pt0(A/e)$ when $pt1(e)=1$ leads to inconsistency, a diachronic incoherence which is rationally unacceptable. According to Lewis and Teller, if an agent fails to update in accordance with Bayesian conditionalization, that is if $pt1(A)=x\neq y=pt0(A/e)$ and $pt1(e)=1$, the following sort of Dutch Book strategy (here simplified) shows that her approach to degree of belief revision is irrational:

> 1.The bookie offers these bets at t0:
> I. A bet on A, conditional on e happening which pays 1
>   costs y
> II. A bet on e which pays y-x
>   costs $(y-x)pt0(e)$
> 2a. At t1, if -e happens, bet I is off, and II is lost
>   NET LOSS= $(y-x)pt0(e)$
> 2b. At t1, if e happens, the bookie buys the bet on A at the current price which the agent judges to be fair:
> III. A bet on A which pays 1
>   costs x
>   NET LOSS= $y + (y-x)pt0(e) - (y-x) - x = (y-x)pt0(e)$.

The agent's conception of what the probability of A is if e happens changes after e happens; consequently, her determination of the fair price for a bet on A conditional on e happening, which becomes simply a bet on A once e happens, alters in a way which allows the bookie to create a Dutch Book against her. Her assessment of the probability at t0 of A if e happens is diachronically incoherent with her assessment of the probability at t1 of A once e happens.

But why, Howson and Urbach would urge us to ask, should this kind of "incoherence" count as a failure of rationality? The agent is not in the sort of structurally problematic belief state characteristic of synchronic incoherence where one essentially has to accept that a set of bets is unfair, even though one judges simultaneously that each individual bet is completely fair: by t1, she no longer likes the price she paid for bet I at t0. The occurrence of e moves her to change her mind, for whatever reason, about the probability of A if e happens, and what is so irrational about that?

As Howson and Urbach stress, there is nothing within the probability calculus itself that says that we must update our probability function in accordance with the standard rule of Bayesian conditionalization, for there is nothing within the probability calculus that says that our value for p(A/e) has to, on pain of irrationality, remain static with the acceptance of e. If $pt1(A/e) \neq pt0(A/e)$ when $pt0(e) \neq 1$ and $pt1(e)=1$, then $pt1(A)=pt1(A/e) \neq pt0(A/e)$. At best the probability calculus tells us what assignments we have to make if all our probabilities for propositions conditional on the evidence remain unchanged after the evidence is accepted.[4] If $pt0(e)<1$, $pt1(e)=1$, and $pt0(A/e)=pt1(A/e)$, then $pt1(A)=pt1(A/e)=pt0(A/e)$ (if my value for p(A/e) does not change when I accept e, then my new value for p(A) is my prior value for p(A/e)). The coherence at issue here is basically synchronic, just as, given that I believed p yesterday, "I believe the same things I believed yesterday and I believe p today" refers to my belief state over time but represents synchronic consistency.

There certainly are cases where p(A/e) could change quite reasonably with the acceptance of e, namely when e is of a kind that would indicate that you ought to change p(A/e). Here is an illustration of this point.

Say that at t1, you do not realize that you are being administered a peculiar psychotropic drug X which effects your judgment in limited ways. The drug sometimes makes you believe the opposite of what you would believe if you weren't on the drug, provided such a change would not involve other major alterations in your belief system. At t1 you don't think that you have been given the drug, but you do recall reading in a trustworthy medical journal that drug X is a mild anti-psychotic which is likely to increase blood pressure.

e: You are administered drug X at t1.
A: Your blood pressure is lowered at t1
$pt1(e) < 1$
$pt1(A/e)$ is low

---

4. See Howson and Urbach, p. 104.

At t2, you find out from a reliable source that you were given drug X at t1, but are currently no longer under its influence. Now that the drug is out of your system, you recall reading in the trustworthy medical journal that drug X is a mild psychosis-*inducer* which is likely to *lower* blood pressure. You are convinced that your previous judgment that p(A/e) is high (which was made while you were on drug X) must be discarded in favor of your new judgment that p(A/e) is low.

$pt2(e)=1$
$pt2(A/e)$ is high $p(A) = pt2(A/e) \neq pt1(A/e)$

But what if our values for the relevant conditionals change when we accept e, where e is *not* of a kind that straightforwardly indicates that we ought to change p(A/e)? In such cases it is difficult to see why our evolving belief state must involve an epistemically problematic incoherence or structural inconsistency. Perhaps we are simply switching from one theoretical framework to another equally acceptable one in a way which we could not have entirely anticipated beforehand. Even van Fraassen ought to find sympathy with this particular point since he believes that the option of complete conversion entailing a new probability function is always a rationally permitted alternative to revising one's prior degrees of belief in accordance with Bayes' theorem, provided one's new theory is no less empirically adequate than one's old theory.[5]

We have identified some important situations where "diachronic incoherence" appears to be perfectly reasonable: the sort of structural inconsistency in the agent's belief state which we find in cases of synchronic coherence is altogether absent. Consequently, we cannot simply assume, as many philosophers do, that diachronic Dutch Books

---

5. It is also worth noting that what initially looks like an unBayesian change in degree of belief could really involve the acceptance of constraints on belief revision beyond the "raw evidence" e. For an explanationist, for example, which available theory possesses the highest degree of explanatoriness in light of what actually happens could be as important a constraint as e itself. (In this case, once e is accepted, pnew(A1) would be based the previous probability value for A1 conditional on both the raw evidence and the information of which available theory possesses the highest degree of explanatoriness in light of what actually happened.) See ch. III for a fuller discussion of this possibility.

are as relevant to questions of rationality as synchronic ones. Hasn't the burden of proof then shifted to defenders of diachronic Dutch Book arguments to provide us with compelling, philosophical reasons for thinking that, these situations notwithstanding, diachronic incoherence, i.e. vulnerability to diachronic Dutch Books, is epistemically problematic? This is a challenge which these defenders, including van Fraassen, have not taken up.

To add further credence to the view that such a challenge cannot be met, we shall now explore some extremely counterintuitive constraints on degree of belief updating rules and assignments which are derivable from the assumption that rational agents must be diachroncially coherent. We first turn to van Fraassen's diachronic Dutch book argument for the principle of reflection in an effort to determine whether the implausibility of its conclusion leads to a *reductio ad absurdum* of van Fraassen's assumption that diachronic Dutch books reveal failures of rationality.

II. THE PRINCIPLE OF REFLECTION

In "Belief and the Will" van Fraassen presents a case for thinking that our current subjective probability for a proposition, conditional on our assigning the proposition a certain value at a time in the future, has to equal that value: $p(A/pt(A)=r)=r$. He calls this constraint on our subjective probability assignments the "principle of reflection."

As van Fraassen notes, adhering to reflection doesn't prevent us from questioning our accuracy as estimators of objective probability. We might realize that we won't possess an understanding of many relevant causal factors at t and conclude that our subjective probability assignment for A at t could diverge sharply from the objective probability of A (we might then set $p(Pt(A)=r/pt(A)=r)$ very low, where P stands for objective chance).

Reflection may allow a healthy skepticism about the extent to which our subjective probability assignments align with objective chance, but it does prohibit us from casting doubt upon our future capacity to make degree of belief judgments in a way that we would honor at present.

Though he presents some more informal considerations in

favor of the principle (which we shall consider in a moment), van Fraassen's most elaborate argument for regarding reflection as a constraint of rationality takes the form of a diachronic Dutch Book argument: he demonstrates how resisting reflection in a particular instance makes one vulnerable to a Dutch Book strategy (implicitly leaving it as an exercise for the reader to show how a more general argument can be built up from the basic mathematical structure of the bets involved, a structure adapted from Teller and Lewis' Dutch Book argument for conditionalization). Here is van Fraassen's example:[6]

A bookie approaches me, offering bets concerning a horse race which will take place tomorrow. All the bets involve one or both of the following propositions:

H: The horse Table Hands will win the race.

E: My probability for H tomorrow morning (t) will be 1/3.

My current probability for E happening, $p(E)$, is 4/10.

Here are the steps in the strategy the bookie devises against me:

1. Since I don't think that I am a perfect handicapper of horses, I don't say $p(H/E)= p(H/pt(H)=1/3)=1/3$, but rather $p(H/E)=p(H/pt(H)=1/3)=1/4$. (My judgment here fails to accord with reflection.)

$p(-H/E)=3/4$ $p(-H \ \& \ E)=p(-H/E)p(E)=3/4(4/10)=3/10$

so the bookie offers me

I. a bet on (-H & E) which pays 1

   costs $p(-H \ \& \ E)=3/10$

2. The bookie want to sell me a bet on -E which

   pays X=$\underline{p(-H \ \& \ E)}$ costs $p(-E)X$
           $p(E)$

so he offers me

II. a bet on -E which pays 3/10(10/4)=3/4

   costs 6/10(3/4)=9/20

3. The bookie wants to sell me a bet on E which

   pays Y= X - the probability -H will have for me
           if E becomes true

6. Readers who are not interested in the details of this example can skip to the bottom of p. xxx.

costs p(E)Y,

so he offers me

III. a bet on E which pays (3/4)-(2/3)=1/12

   costs 4/10(1/12)=1/30

Total cost for I & II & III= 3/10+9/20+1/30=47/60

4. Here is summary of my wins & losses tomorrow:

   If -E happens (pt(H)≠1/3), I win II and lose I & III

   net loss= total cost of bets - winnings=

   47/60 - 1/12 = 1/30

   If E happens (pt(H)=1/3), I have so far won III & lost II

   loss so far= 47/60 - 1/12 = 42/60

Now, since I think the probability of H is 1/3, I should be willing to sell my bet on -H & E (which, since E has happened, is basically a bet on -H with payoff 1) for 2/3.

   net loss= 47/60 - 2/3 = 1/30

I sustain a loss of 1/30, come what may. By not subscribing to reflection when I determined my personal probability for H & -E, I made myself vulnerable to a Dutch Book strategy. Disobeying reflection leads to incoherence, the argument goes, so the principle of reflection must be seen as a "new requirement of rationality."

But is there any natural appeal to this requirement? Does the conclusion of van Fraassen's argument accord well with our intuitions about what should and shouldn't count as a constraint of rationality, or does it clash with them in a way which would vindicate our suspicions that diachronic Dutch Books have no interesting epistemic implications?

In what we may regard as an effort to settle this issue in his favor, van Fraassen rather cryptically tries to explain how a voluntarist conception of the nature of epistemic judgment makes the principle attractive. According to a voluntarist view, a voiced subjective probability judgment is not a mere statement of autobiographical fact, but rather the expression of an intention to take a stand, to make a commitment. Presumably the force of this idea is that the subjective probability judgments we come to see as our own attain this status by an act of will, the result of deliberate decision. They are not opinions which are forced upon us by inflexible rules applied to evidence, nor are they neutral claims about our state of mind without any essential connection to intentions or actions.

Just how van Fraassen uses these considerations to create a tenable interpretation of reflection is rather unclear, but he seems to suggest that seeing a degree of belief avowal as the expression of a commitment enables one to draw a helpful analogy with promise keeping. According to van Fraassen (or so I gather), openly questioning one's capacity to make personal probability judgments in the future in a way one could currently respect would undermine the meaningfulness of one's degree of belief avowals, much the same way that expressing doubts about one's ability to carry through a promise made in the future would undermine the meaningfulness of the act of making the promise. By violating reflection, one would be indicating beforehand an inability to follow through with the kind of commitment entailed by sincerely expressing degree of belief judgments. A listener would have to suspect that I at some point display a lack of integrity either at present by being insincere in my claims about the future, or (if my listener believes me now) in the future by giving voice to a commitment I have already prohibited my listener from taking seriously.

> I must stand by my own cognitive engagement as much as I must stand by my own expressions of commitment of any sort... I can no more say that I regard A as unlikely on the supposition that tomorrow morning I shall express my high expectation of A, than I can today make the same statement on the supposition that tomorrow morning I shall promise to bring it about that A. To do so would mean that I am now less than fully committed (a) to giving due respect to the felicity conditions for this act, or (b) to standing by the commitments I shall overtly enter.(van Fraassen 1984, p.255)

It might be appropriate to say that if I fail to obey *synchronic* reflection [p(A/p(A)=r)=r] I would sabotage my capacity to count myself as being the measure of my own degrees of belief, let alone my capacity to follow through with the commitments entailed by my subjective probability judgments. Not assigning r to the conditional $p(A/p(A)=r)$ would undermine the meaningfulness of my degree of belief avowals. "My current degree of belief for A, on the supposition that I have current degree of belief r for A, does not equal r, but rather equals s" sounds quite paradoxical; a listener would have to conclude that no commitment to a value for the degree of belief in A has been made. Showing that there might be reason for viewing violations of diachronic reflection as somehow self-defeating in the fashion of the promise keeping example is much less straightforward.

Right at the start one can question whether it is always self-defeating to say that A is unlikely on the supposition that I promise to bring about A tomorrow. Again, the synchronic violation is less controversial. If I say that I sincerely promise you A (not just that I promise to try to bring about A), but A is unlikely, it really looks like my promise can't be taken very seriously. A problem arises for van Fraassen's claim if we insist that it can be appropriate to say "A is unlikely on the supposition that I promise A tomorrow" when my not being able to carry through with the promise I make tomorrow is conditional on some highly likely, but not inevitable circumstances C beyond my control (such as the development of a particular kind of unfortunate complication during brain surgery). Consider the following example.

M: I promise A at t1 (tomorrow).
$p(A/M\&C)=p_c(A/M)=1/10$ (It's unlikely that I'll carry through with promises I make at t1 if C.)
$p(A/M\ \&\text{-}C)=p\text{-}c(A/M)=9/10$ (It's likely that I'll carry through with promises I make at t1 if -C.)
$p(C)=9/10$ (C is likely)
By the theorem of total probability:
$P(A/M)=p_c(A/M)p(C) + p\text{-}c(A/M)p(\text{-}C)=9/100 + 9/100=18/100$

A is unlikely on the supposition that I promise A tomorrow. Provided I inform my listener that $p(C)$ is high, $p_c(A/M)$ is low, and $p\text{-}c(A/M)$ is high (justifying my claim that $p(A/M)$ is low), if -C happens, I at no point sacrifice my integrity; my commitments can be taken seriously. If C happens, it's a good thing that I gave warning about my future inability to carry through with commitments and distanced myself from my potential future self: that future self's rationality or knowledge has been diminished. Given the circumstances, having admitted that A is unlikely on the supposition that I promise A tomorrow was entirely appropriate.

We can draw upon this example to illustrate that there are plenty of situations where disobeying reflection seems to be quite reasonable. It looks like it is possible to say "I regard A as unlikely on the supposition that tomorrow I shall express my high expectation of A" without at any point appearing to sacrifice my integrity or status as a rational agent whose avowals can be taken seriously, provided circumstances conspire in my favor.[7] Substituting "I express my high expectation of A at t1" for M in the promise keeping example, we can devise an objection aimed directly at the principle of reflection. (Again, C could stand for one of any number of unpleasant, mind-altering calamities: the onset of a devastating senility, being administered a powerful psychotropic drug, becoming the guinea pig in a bizarre set of neurological experiments, etc.)

At present (t0), I'm in little position to make a judgment about A, but I know that tomorrow (t1) I'll be exposed to a great deal of relevant evidence. I make the following judgments:

$$p_c(A/M) = 1/10 \quad p\text{-}c(A/M) = 9/10 \quad p(C) = 9/10$$
$$p(A/M) = p_c(A/M)p(C) + p\text{-}c(A/M)p(\text{-}C) = 18/100$$

If I have made it clear to my listeners that $p_c(A/M)$ is low, $p\text{-}c(A/M)$ is high, and $p(C)$ is high, I can say $p(A/M)$ is low (violating reflection) and still, provided -C occurs tomorrow, have the commitments I make by my degree of belief avowals at t1 taken seriously.

Even if the feared conditions do come to pass, why should a sincere past expression of a doubt I no longer acknowledge—a doubt conveyed by a past self I am dissociated from—count as self-defeating? It seems that I am not inhibited at that point in the future from fully committing myself to the degree of belief judg-

---

7. My remarks here accord very well with Christensen's views in "Clever Bookies & Coherent Beliefs."

ments I make at t1. One could note here that I am not fully rational, but the root of the irrationality doesn't seem to lie in my disobeying reflection. (Indeed, as Christensen notes in "Clever Bookies and Coherent Beliefs," in the face of a potential irrationality inducing C, violating reflection appears to be the only rational option.) We have shown above how I can violate reflection without at any point being unable to have my commitments taken seriously or falling prey to the irrationality which could have been (but wasn't) induced. Furthermore, a violation on the part of an otherwise rational agent seems in general only to indicate that the agent believes that some kind of change could with noticeable likelihood occur which would dissociate her future self (with its ways of responding to evidence) from her present self.

We have focused up to now on cases where C induces a transition to a more irrational or ignorant state, but the change doesn't have to be understood in this way for the objection to work. C could stand for some kind of conversion experience (not even necessarily a religious one; perhaps one could think of a conversion in science or philosophy). In these cases, why would the assignment for the conditional $p(A/pt(A)=r)$ have to respect the value at t? There is nothing about the nature of epistemic judgments per se which persuades me that the following intuitively attractive way of making the conditional assignments in potential future conversion cases is rationally unacceptable: $p(A/pt(A)=r \ \& \ C)\neq r$, $p(A/pt(A)=r \ \& \ -C)=r$, $0<p(C)<1$, so $p(A/pt(A)=r)\neq r$.

The principle of reflection is equally dubious when we take A to be a particularly outlandish proposition: the world began with my birth; I can fly like a bird; my cat will be the next Messiah. As Christensen notes, when A is patently absurd, my current personal probability for A, conditional on my according $p(A)$ a high value in the future, should remain extremely low. I know now that if I ever came to believe A, I would have entered an exceptionally irrational state of mind; consequently, I have good reason not to "stand by" such future commitments by honoring reflection.

The unreasonableness of the principle of reflection in this sort of case and the others we have considered casts serious doubt upon the legitimacy of the mode of argument van Fraassen uses to defend the principle. We can draw a similar conclusion for the temporally extended principle of reflection, which is also derivable from a diachronic Dutch Book argument.

III. THE TEMPORALLY EXTENDED PRINCIPLE OF REFLECTION

Since van Fraassen is a firm believer in the power of diachronic Dutch Book strategies, he must accept the following constraint on reflection which I shall call the "temporally extended principle of reflection": the value of $p(A/pt1(A)=a1...\& \ ptn(A)=an))$ must equal the value of the temporally last $pti(A)$ $(ptn(A))$. Here is a demonstration that violating this principle makes an agent vulnerable to a diachronic Dutch Book strategy.

Consider the probability of A, conditional on A being assigned different probability values at later times:

$p(A/(pt1(A)=a1 \ \& \ pt2(A)=a2 \ ... \ \& \ ptn(A)=an))$.

If an agent does not always take the value of the temporally last $pti(A)$ as the value for the conditional probability, a Dutch Book can be created against her. For the sake of simplicity, we shall look at A conditional on A being assigned probability values at two later times, t1 and t2, where the sets of possible values for $pt1(A)$ and $pt2(A)$ form finite partitions:

$S=\{si: pt1(A)=si;i=1...n\} \ R=\{rj: pt2(A)=rj;j=1...m\}$.

We assume that A is a proposition of a type that cannot be accorded probability 0 in the time frame we are considering.

Corresponding to each possible pair, (si,rj), there is a value of the function $f(si,rj)= pt0(A/(pt1(A)=si \ \& \ pt2(A)=rj))$.

---

8. Readers who are not interested in the technical details of this rather involved argument may skip to p. xxx.

At t0 we could compile a chart showing all the values for this function:

```
       r1   r2....rm
s1    f11  f12  f1m
s2    f21  f22  f2m
 .
 .
 .
sn    fn1  fn2  fnm
```

Henceforth, (sa,ra) shall refer to the pair of variables 'the actual values for (pt1(A),pt2(A)), whatever they turn out to be.' Once the values satisfying (sa,ra) (say (xs,xr)) become known, it would be possible to look back at our chart to determine the value of f(xs,xr).

We are considering the case where an agent does not subscribe to a rule whereby ∀si,rj f(si,rj)=rj.

∃si,rj f(si,rj)≠rj, so pt0(f(sa,ra)≠ra)≠0

(pt0[pt0(A/(pt1(A)=sa & pt2(A)=ra)≠ra))]≠0).

Here is the bookie's strategy:

At t0

1. I. Offer a bet on pt0(f(sa,ra))≠ra. (As we noted , the agent assigns a non zero probability, say d, to the proposition at hand.)

Determine the payoff for the bet as follows: find the minimum value m for |f(si,rj)-rj| using the chart above.

cost of bet I: dm

payoff: m

2. For all pairs (si,rj) such that f(si,rj)_rj, offer either IIa or IIb, depending on the value of f(si,rj):

If f(si,rj) > rj, offer

IIa. a bet on A, conditional on pt1(A)=si and pt2(A)= rj((sa,ra)=(si,rj)).

The payoff is w, where w is some amount greater than 1. Since pt0(A/(pt1(A)=si & pt2(A)=rj)) = f(si,rj), the cost of each IIa bet is wf(si,rj).

If f(si,rj) < rj, offer

IIb. a bet on -A, conditional on pt1(A)=si and pt2(A)=rj.

$((sa,ra)=(si,rj))$.

The payoff is w, where w is some amount greater than 1.

Since $pt0(-A/(pt1(A)=si$ & $pt2(A)=rj))$ =

$1 - pt0(A/(pt1(A)=si$ & $pt2(A)=rj))$ =

$1 - f(si,rj)$,

the cost of each IIb bet is $w(1-f(si,rj))$.

At t2

The values satisfying (sa,ra)(say (xs,xr)) have become known. If $f(xs,xr)=xr$,

bet I is lost

All bets in IIa & IIb are called off. (The bets in 2 correspond only to the (si,rj) pairs where $f(si,rj)≠rj$. Each II bet is conditional on (sa,ra) being equal to an (si,rj) pair, where $f(si,rj)≠rj$.)

Net loss for the agent = dm STOP

If $f(xs,xr)≠xr$,

The bookie offers either IIIa or IIIb, depending on the value of $f(xs,xr)$.

If $f(xs,xr) > xr$, offer

IIIa. A bet on -A.

The payoff is w (as in IIa).

Since $p(-A)= 1 - pt2(A) = 1 - xr$,

the cost of IIIa is $w(1-xr)$.

If $f(xs,xr) < xr$, offer

IIIb. A bet on A.

The payoff is w (as in IIb).

Since $p(A)=pt2(A)=xr$,

the cost of IIIb is wxr.

Now, in either case ($f(xs,xr)>xr$ or $f(xs,xr)<xr$)), a net loss for the agent results, independent of whether or not A obtains.

If $f(xs,xr) > xr$:

Total bet costs= cost of I + the cost of the IIa bet conditional on $(sa,ra)=(xs,xr)$ + the cost of IIIa. (All other II bets are called off since they are conditional on (sa,ra) being equal to other (si,rj) pairs.)

Total bet costs= dm + wf(xs,xr) + w(1-xr)).

The agent has won bet I with payoff m.

The agent will also win either bet IIa (if A happens) or bet IIIa (if -A happens). Since the payoff is w in either case, the agent will win w.

Total win = m + w.

Net loss   = total bet costs - total win

= dm + wf(xs,xr) + w(1-xr) - m - w

= dm + w(f(xs,xr)-xr)) - m.

Since m is the minimum value for |f(si,rj)-rj|, m ≤ (f(xs,xr)-xr)). w > 1, so

[w(f(xs,xr)-xr) - m] > 0.

The agent sustains a net loss. STOP

If f(xs,xr) < xr:

Total bet costs= cost of I + the cost of the IIb

bet conditional on (sa,ra)=(xs,xr) + the cost of IIIb.

(All other II bets are called off since they are conditional on (sa,ra) being equal to other (si,rj) pairs.)

Total bet costs= dm + w(1-f(xs,xr)) + wxr.

The agent has won bet I with payoff m.

The agent will also win either bet IIb (if -A happens) or bet IIIb (if A happens). Since the payoff is w in either case, the agent will win w.

Total win = m + w.

Net loss   = total bet costs - total win

= dm + w(1-f(xs,xr)) + wxr - m - w

= dm + w(xr - f(xs,xr)) - m.

Since m is the minimum value for |f(si,rj)-rj|,

m ≤ (xr - f(xs,xr)). w > 1, so

[w(xr - f(xs,xr)) - m] > 0.

The agent sustains a net loss. STOP

The agent suffers a net loss, come what may: f(xs,xr)=xr, f(xs,xr)>xr, and f(xs,xr)<xr. By judging that ∃si,rj f(si,rj)≠rj, and hence ascribing a nonzero value to pt0(f(sa,ra)≠ra), she makes herself vulnerable to a Dutch Book strategy. If she had decided that ∀si,rj f(si,rj)=rj, this strategy couldn't have been used. By judging pt0(f(sa,ra)≠ra)=0, she would have been unable to accept bet I.

A Dutch Book argument following the basic strategy above could be developed for any n where the agent judges that it is not always the case that $p(A/(pt1(A)=a1...\& \; ptn(A)=an))=an$. Since van Fraassen believes that diachronic Dutch Book arguments are effective in pointing out principles of rationality, he would have to accept the constraint that the value of $p(A/pt1(A)=a1...\& \; ptn(A)=an))$ must equal the value of the temporally last $pti(A)$ $(ptn(A))$.

There seems to be little intuitive appeal to this constraint. One could say, perhaps, that it would be applicable to the probability assignments of an ideally rational agent who possessed the foreknowledge that 1. her capacity to make what she would presently regard as reasonable judgments would not be compromised, and 2. she would not undergo radical, arational conversion (a conversion unexplainable by mere reaction to evidence, and not based on any rational justification). The judgment of the latest time $ptn(A)=an$ would, presumably, have the advantage of being made in light of the most exposure to evidence and reasoning. It would incorporate all the information and insight used to make previous $pti(A)$ judgments, and then some.

According to van Fraassen, however, principles of rationality disclosed by means of diachronic Dutch Book arguments are supposed to count as constraints on *our* epistemic judgments, not just the epistemic judgments of ideally rational agents with superhuman foreknowledge. Given that we are vulnerable beings whose capacities for making judgments in a way we presently regard as reasonable can erode through time, taking the last $pti(A)$ as the value for the conditional in question doesn't always seem to be the best choice. Here are two illustrations:

1. Suppose that I suspect that it's likely that I shall become senile in my later years. If I now conditionalize A on my believing A to different degrees throughout my life (up to tn, when I shall be very old), it would seem to be most reasonable for me to place less than full weight on the judgment $ptn(A)=an$ as I establish a value for $pt0(A/pt1(A)=an...\&ptn(A)=an)$. The passage of time may expose me to more evidence and insight, but it could also lead me

to mental deterioration.

2. Suppose that I acknowledge that it is possible that arational influences (such as emotional needs, for example) might, at several points in my life, cause my future self to adopt a preference for alternative world pictures which would strike me now (if I had access to my future evidence and insight) as just as rational as the world picture I possess before each conversion. This is not to say I would be committed to the view that equally empirically adequate theories are equally reasonable or rational, but rather that in some cases, our best reasoning leaves more than one reasonable option open. One might wonder if it would be more philosophically appropriate to replace the commitments and leaps of faith involved in such a position with agnosticism, but it seems safe to say that there is nothing incoherent about these conversions. At any rate, being a constructive empiricism, van Fraassen wouldn't raise any objections to this scenario.

If I suspect that I might convert several times from t1 to tn, it seems arbitrary for me to choose my judgment $ptn(A)$ as the value for $pt0(A/(pt1(A)=a1...\&ptn(A)=an))$. $ptn(A)$ might end up differing from $ptn-1(A)$, not on the basis of new evidence, but rather as the result of an arational change in world picture. When I establish the value for the conditional at t0, why should I honor my judgment at tn at the expense of my judgment at tn-1?

As in the case of the original principle of reflection, violating the temporally extended principle of reflection appears to be perfectly reasonable. We now turn to another class of degree of belief assignments which result in diachronic incoherence, yet seem consistent with the norms of epistemic rationality. This kind of case is less straightforward and somewhat harder to interpret than the previous ones, but perhaps it will still be able to provide some further support for the view that diachronic Dutch Book arguments cannot be used to highlight failures of rationality.

---

9. There is no implication here that A occurs at the same time as B: the agent is assigning a value to $p(A/B)$.

IV. A PROHIBITION AGAINST ASSIGNING A PROBABILITY VALUE TO SOME SPECIAL CONDITIONAL PROPOSITIONS

Say an agent assigns probability x to the proposition When B happens, A happens.[9] She is asked what value she would assign to the proposition When C happens, and B happens, A happens, where C satisfies the following conditions:

1. C is causally irrelevant to A, and, taken on its own, statistically irrelevant to A.
2. The agent has no more reason to think that C will raise the probability of A, given B than that C will lower the probability of A, given B.

Because C satisfies these conditions, the agent decides that her personal probability for *When C happens, and B happens, A happens* is the same as the probability she assigned to *When B happens, A happens*, namely x: p(A/B&C) = pC(A/B) = p(A/B)= x.[10] Consequently, she will consider a fair price for a "dual conditional" bet on A—a conditional bet on A if B happens, conditional on C happening—with payoff 1 to be x, the same cost as a simple conditional bet on A if B happens with payoff 1.

The agent appears to be using perfectly good probabilistic reasoning in setting her personal probability values here, but she will run into diachronic Dutch Book trouble when C is a proposition about the bookie's betting behavior which satisfies the conditions 1 & 2 above. Here is a demonstration of how her approach to assigning degree of belief values leads to diachronic incoherence, even when she follows probabilistic updating rules (such as normal conditionalization) perfectly.[11]

Consider possible future events A1, A2 and B such that:

e lowers p(-A1/B)	[pe(-A1/B)<p(-A1/B)]
(raises p(A1/B))	
-e lowers p(-A2/B)	[p-e(-A2/B)<p(-A2/B)]

---

10. To clarify the symbolism here: pC(W) = p(W/C).

11. This argument is rather complicated (requiring some tricks from the kind of Dutch Book argument van Fraassen presents in "Belief and the Will," plus a few others), so the reader may wish to skip to p. xxx to avoid all the technical details.

(raises p(A2/B))

A1 or -A1, and A2 or -A2 will happen after time t; e or -e, and B or -B will happen before time t. They are all ordinary events in the world which are uninfluenced by human actions (we can imagine, for example that they all have to do with the weather).

I. The bookie offers these conditional bets:

a. A bet on -A1 if B happens, conditional on C1: The bookie offers no unconditional bets on A2 by t.

b. A bet on -A2 if B happens, conditional on C2: The bookie offers no unconditional bets on A1 by t.

Now C1 and C2 satisfy conditions 1 & 2 above. C1, for example, is causally irrelevant to -A1 (-A1 is uninfluenced by human actions), and taken on its own, C1 is statistically irrelevant to -A1 (the bookie has to decide to offer no unconditional bets on A2 *before* A1 or -A1 happens in order to make C1 true). In addition, the agent has no more reason to think that C1 will raise the probability of -A1, given B than that C1 will lower the probability of -A1, given B. Consequently, the agent will calculate the fair prices for bets Ia & Ib by setting them equal to the fair prices for simple conditional bets (with the same payoffs) on -A1, given B and on -A2, given B, respectively.

The costs & payoffs of Ia & Ib are as follows:

a. We break down bet Ia into two parts:

Conditional on C1:

	costs	pays
i. bet on -A1 & B	p(-A1 & B)	1
ii. bet on -B	p(-A1/B)p(-B)	p(-A1/B)

[i+ii is essentially a bet on -A1 if B happens (which pays 1):

total cost= p(-A & B) + p(-A1/B)p(-B) =

p(-A1/B)p(B) + p(-A1/B)p(-B) =

p(-A1/B)[p(B) + p(-B)] =

p(-A1/B)

If -B happens, the agent just gets paid back the cost of the bets: the conditional bet is off resulting in no gain, no loss.]

b. We also break down bet 1b into two parts:

Conditional on C2:

	costs	pays
i. bet on -A2 & B	p(-A2 & B)	1
ii. bet on -B	p(-A2/B)p(-B)	p(-A2/B)

The bookie waits for e or -e to occur, and offers different bets depending on what happens:

If <u>e</u>   If <u>-e</u>

↘   ↘

II. Offer bet on B   **II. Offer bet on B**

costs: p(B)[p(-A1/B) - pe(-A1/B)]   costs: p(B)[p(-A2/B) - p-e(-A2/B)]

pays: p(-A1/B) - pe(-A1/B)   pays: p(-A2/B) - p-e(-A2/B)

The bookie waits for it to be known whether B or -B

If B	If -B		If B	If -B
	↘			↘
	bet I. off			bet I. off
	II. lost			II. lost
	net loss			net loss
	for agent			for agent
	STOP			STOP

↘   ↘

III. Offer bet   III. Offer bet
on A1   on A2

costs	pays		costs	pays
pe(A1/B)	1		p-e(A2/B)	1

Time t passes

bet Ia on, bet Ib off   bet Ia off, bet Ib on

Game over

No matter what happens, the agent sustains a net loss.

We have seen that a net loss occurs if -B takes place (totaling the cost of II).

If B happens, the agent loses money, independent of whether Ai or -Ai occurs. Without loss of generality, we can show that this is true in the left hand case: e occurs and an unconditional bet on A1 is offered. The structure of the bets if -e occurs and an unconditional bet on A2 is offered is the same. (Recall the bets on the left are offered only if e happens, and the bets on the right are offered only if -e happens.)

TOTAL COST OF BETS I & II =

$p(-A1 \& B) + p(-A1/B)p(-B) + p(B)[p(-A1/B) - pe(-A1/B)$

$= p(-A1 \& B) + p(-A1/B)[p(-B) + p(B)] - p(B)pe(-A1/B)$

$= p(-A1 \& B) + p(-A1/B) - p(B)pe(-A1/B)$

Since we are considering the case where B occurs, we subtract from the total cost the amount won for bet II to find the total loss so far (before bet III is offered).

total cost   payoff from II

$p(-A1 \& B) + p(-A1/B) - p(B)pe(-A1/B) - [p(-A1/B) - pe(-A1/B)]$

$= p(-A1 \& B) + p(-A1/B) - p(B)pe(-A1/B) - p(-A1/B) + pe(-A1/B)$

$= p(-A1 \& B) + pe(-A1/B)[1-p(B)]$

Now we add the cost of bet III to the total loss before bet III was offered.

previous loss   cost of III

$= p(-A1 \& B) + pe(-A1/B)[1-p(B)] + pe(A1/B)$

$= p(-A1 \& B) + [pe(-A1/B) + pe(A1/B)] - p(B)pe(-A1/B)$

$= p(-A1 \& B) + 1 - p(B)pe(-A1/B)$

The agent wins 1 (and only 1) whether or not A1 happens (the conditions for bet Ia counting have been met so the agent wins either Ia:the bet on -A1 paying 1 or III: the bet on A1 paying 1).

Winning 1 is not enough to offset a net loss:

previous total loss   latest win

$[1 + p(-A1 \& B) - p(B)pe(-A1/B)] -$ 1

$= p(-A1 \& B) - p(B)pe(-A1/B)$

$= p(B)p(-A1/B) - p(B)pe(-A1/B)$

$= p(B)[p(-A1/B) - pe(-A1/B)]$

NET LOSS

As stipulated at the beginning of the example, $pe(-A1/B)<p(-A1/B)$, so the net loss is greater than 0 (whether or not A1 happens).

Even though the agent's probabilistic reasoning appears perfectly in order, she winds up the victim of a diachronic Dutch

Book. A major culprit is, of course, the presence of the dual conditional bets Ia and Ib, both of which are conditional on a proposition about the bookie's betting behavior. Such propositions, in other contexts, can lead rather quickly to Dutch Book problems, as the following simple examples demonstrate.

We see right away that it would be problematic (at least from a practical standpoint) for an agent to buy unconditional bets on propositions concerning the bookie's behavior. Consider the following strategy concerning proposition K: *The bookie offers the agent a bet at t1.*

If the agent sets $pt0(K)=0$:

The bookie offers a bet on -K at t0 with payoff W for $W(pt0(-K)=1)$. At t1, the bookie then offers a bet on any arbitrary proposition with a payoff less than W (so the agent loses the bet on -K). Even if the agent goes on to win the second bet, the gain from the win would be offset by the cost of the previous lost bet (W), resulting in a net loss.

If the agent sets $pt0(K)\neq0$ (say $pt0(K)=r$):

The bookie offers a bet on K with payoff W for $p(K)W$. At t1, the bookie then refrains from offering further bets. The agent loses the bet on K, ending up with a net loss.

The agent sustains a loss, come what may.

A simple scenario can also be devised to show how devastating bets conditional on the bookie's behavior can be.

At t0

1. The bookie offers two bets:

a. A bet on A, conditional on O1: *The bookie offers a bet on B at t2.*

b. A bet on -A, conditional on O2: *The bookie offers a bet on D at t2.*

(A or -A happens at t1.)

At t2

If A has happened, the bookie offers only a bet on D with such a low cost/low gain that even if the agent were to win the bet, the gain would be too small to offset the loss of bet 1b. Since bet 1a is called off (no bet on B is offered), the agent sustains a net loss.

If -A has happened, the bookie offers only a bet on B with such a low cost/low gain that even if the agent were to win the bet, the gain would be too small to offset the loss of bet 1a. Since bet 1b has been called off (no bet on D is offered), the agent sustains a net loss.

What are the implications of these simple examples for the issue at hand? Do they lend further support to the view that diachronic incoherence is not necessarily epistemically problematic, or are they irrelevant? Do they perhaps undermine the force of the case we considered at the beginning of the section by suggesting that assigning probabilities to conditional propositions involving the bookie's betting behavior is in itself irrational? If this were so, our original example could not then be put forward as evidence that diachronic Dutch Books fail to reveal breaches of rationality.

Under one possible interpretation, the simple cases involve diachronic incoherence. If this interpretation is the correct one, it would be question-begging for the Dutch Book advocate to say at this juncture that assigning a probability to propositions involving the bookie's betting behavior is clearly irrational because the agents in the simple cases wind up with Dutch Books. Given the absence of a general account of why diachronic incoherence is epistemically problematic, she could only support her cause in the current context by presenting independent reasons for thinking that such probability assignments are irrational. But what could such reasons be? Here my imagination is at a loss; I cannot even begin to picture how someone could answer this question (would she try appealing to a peculiar, bad argument about the dangers of self-reference?) The clear practical disadvantage to assigning a degree of belief to these sorts of events is of course of no use in highlighting any epistemic flaws. If no further response is available, we might be tempted to turn the tables on our diachronic Dutch Book advocate and insist that the simple cases—provided that they do genuinely involve diachronic incoherence—show all the more that diachronic incoherence is not an epistemic problem.

According to another possible interpretation, the simple cases really involve only synchronic incoherence. The agent is perhaps in a position where she simultaneously has to regard the bets she accepts as fair under one criterion (the prices are based on her current degrees of belief), but unfair under another (she should know that as soon as her probability assignments for the relevant propositions become determinate, the bookie will have the advantage of being able to use this information to influence his future choices and make her lose). If we were to take this line, we would have reason to be skeptical about the epistemic merits of assigning a probability value to at least some propositions involving the bookie's betting behavior.

We still have to settle whether the conditional propositions in the original, complex example would fall into this category, or whether there are significant differences between these and the conditional propositions from the second simple example which prevent such an association (we leave aside the first simple case since the objects of the bets are so patently dissimilar). It seems to me that there are important, relevant features which distinguish the conditional propositions in the complex case from those in the simple case.

In the complex case, the probability assignments for the dual conditional propositions are determined by appeal to highly general probabilistic considerations which seem to indicate which values are appropriate, independent of what "C" stands for. For the rules to apply, however, C must satisfy two constraints:

1. C is causally irrelevant to A, and, taken on its own, statistically irrelevant to A.

2. The agent has no more reason to think that C will raise the probability of A, given B than that C will lower the probability of A, given B.

If we look back to bets 1a and 1b from the simple example, we see that the propositions about the bookie's betting behavior O1 and O2 do not satisfy the first of these conditions (the second is not relevant since it applies only to dual conditional propositions):

1.a. A bet on A, conditional on O1: *The bookie offers a bet on B at t2.*

1b. A bet on -A, conditional on O2: *The bookie offers a bet on D at t2.*

(A or -A happens at t1.)

O1 and O2 are not statistically irrelevant to A, since the bookie makes O1 or O2 true *after* A or -A happens: he will decide after A or -A occurs which conditional bet shall be called off by his offering only a further bet on D or B, respectively. The probability that A is true, given that O2 is true is extremely high (and certainly is not the same as the probability of A by itself).

The bookie does not have this kind of power to immediately dupe the agent in the complex example precisely because the propositions involving the bookie's betting behavior which are conditionalized on do satisfy condition 1.

Ia. A bet on -A1 if B happens, conditional on C1: *The bookie offers no unconditional bets on A2 by t.*

Ib. A bet on -A2 if B happens, conditional on C2: *The bookie offers no unconditional bets on A1 by t.*

(A1 or -A1, and A2 or -A2 happen after t.)

The bookie has to make C1 or C2 true before A1 or -A1, A2 or -A2 happen.

This fact seems to make the option of regarding the incoherence which surfaces later on as synchronic considerably less appealing than it might be for the simple example. In the case at hand, factors beyond conditionalizing on the bookie's betting behavior are crucial. We see that if the initial constraints of the example weren't satisfied (e lowers $p(-A1/B)$, -e lowers $p(-A2/B)$), running the same strategy could lead the agent to either gain or break even (the total loss if B is $p(B)[p(-Ai/B) - pe(-Ai/B)]$). Leaving aside any consideration of the influence of e, the bookie is powerless to generate a certain loss from Ia, Ib and II. A bet on B.

If we are to accept the incoherence in question as genuinely diachronic, we can reiterate our challenge to the diachronic Dutch Book sympathizer: explain why the agent's degree of belief assignments must be seen as epistemically compromised. The probabilistic reasoning described in the opening paragraphs of the section seems in order, so why should the fact that C1 and C2, which satisfy the relevant constraints, concern the bookie's behavior make this approach epistemically irrational?

Admittedly, the peculiar status of the propositions involved makes the example more difficult to interpret. Even if we were to decide that, because of its inconclusiveness, we ought not depend on it too heavily in our case against regarding diachronic incoherence as irrational, we can still rely on the more decisive victories from our discussions of the principle of reflection, and the temporally extended principle of reflection. The counterintuitive nature of the verdicts provided by the diachronic Dutch Book arguments in these instances should be more than enough of an indication that we shouldn't use diachronic Dutch book arguments to draw epistemic conclusions.

CHAPTER FIVE

Varieties of Explanationism and Fine's Critique

One would be hard pressed to find rhetorical attacks against scientific realism which approach the level of intensity Arthur Fine's have achieved. The tone of these assaults is funereal and merciless: scientific realism seems like a silent corpse or, at best, a tired and pathetic little spook that still tries to zoom around the house, but has long outstayed its welcome.

> Realism is dead... To be sure, some recent philosophical literature has appeared to pump up its ghostly shell and to give it new life. I think these efforts will eventually be seen and understood as the first stage in the process of mourning, the stage of denial. But I think we shall pass through this first stage and into that of acceptance, for realism is well and truly dead, and we have work to get on with, in identifying a suitable successor. (Fine 1984, p. 83)

> Indeed, as we shall see, the realist programme has degenerated by now to the point where it is quite beyond salvage. A token of this degeneration is that there are altogether too many realisms. It is as though by splitting into a confusing array of types and kinds, realism has hoped that some one variety might yet escape extinction. (Fine 1986, p. 149)

Fine overstates the extent of his victory, but we should grant that he has made valuable contributions to the debate over the status of our beliefs about unobservables. His criticisms of some early attempts by Smart and Boyd to forge out an explanationist defense of realism have revealed fundamental problems with certain realist arguments and tactics. Although later developments in Boyd's work handle some of these difficulties fairly well, ultimately Fine's critique is suggestive of shortcomings which will face any abductive argument for realism, and, more significantly, any externalist, naturalist form of explanationism. This consequence is quite serious for Boyd, whose commitment to externalism/reliabilism and naturalism marks what is most distinctive about his version of explanationism.

Since it's not always so clear what the labels amount to, here is a brief overview of how each is relevant to describing Boyd's position:

Naturalism: the view that philosophy is a branch of the natural sciences, and so proceeds by the application of methodological principles which are akin to methodological principles used in more conventional scientific investigations. Any philosophical claim is, essentially, an empirical one, and hence must be defended a posteriori by appeal to empirically substantiated rules of inference and justification; there are no rules, methods, or claims which are acceptable a priori or posses an a priori justification.

Boyd, as a philosophical naturalist, regards scientific realism as an empirically warranted thesis which is established by using some of the basic rules and methods (such as abduction) which are routinely appealed to in the confirmation of scientific findings.

Reliabilism: the view that the most illuminating philosophical analysis of the concepts which form the subject matter of traditional epistemology— such as knowledge, rationality, and justification—invoke the notion of reliable belief-forming processes. Though reliabilists have developed refinements in attempts to handle Gettier-type examples and other objections, a basic reliabilist account of knowledge, for example, identifies knowledge with belief which is true and results from a relevant, reliable belief-forming process. To be rational or justified, a belief need not be true, though it must be arrived at by a reliable belief-forming process.

Externalism: the view that whether a belief counts as rational, justified, or an instance of knowledge depends on the presence of circumstances cognitively inaccessible to the agent. To be rational, for example, a belief does not necessarily have to pass standards for evaluation which the agent can assess; a belief is rational when it results from a reliable belief-forming process, even if the agent has no idea that the relevant process is in fact reliable, and even if, in principle, the agent could never conclusively establish that the relevant process is reliable.

The externalist, reliabilist responses to classic skeptical and empiricist challenges are rather swift and tidy: our ordinary beliefs about the world are rational, despite the difficulties presented by Descartes' demon, because our ordinary beliefs in fact result from reliable belief-forming processes; our inductive inferences are justified despite Hume's problem because our inductive practices conform to reliable belief-forming processes; our projectibility judgments, and hence our abductions, are justified despite Goodman's new riddle of induction because these judgments are in fact based on theoretical commitments drawn from a heritage of approximately true background beliefs. (This latter solution will be particularly relevant when we evaluate Boyd's claim that the anti-realist, unlike the realist, cannot accept even claims about empirical adequacy or instrumental reliability as justified.)

Boyd's naturalized, externalist realism, and non-naturalized internalist realism share the convictions that our norms of inquiry are shaped by our interest in trying to get at the truth (in a way that exceeds what the anti-realist would allow), that there is more to rationality than coherence and empirical adequacy, and that theoretical results in science are, at least at times, rationally compelled, but that is where the similarity ends. The internalist explanationist wants to locate the source of rational constraint within the cognitive grasp of the agent. What narrows down the range of legitimate, rational options among the equally empirically adequate alternatives at a given stage of inquiry is not the fact that one option best conforms to a heritage of true theoretical commitments, or reliable belief-forming practices, but rather that one option best conforms to standards for inquiry which the agent herself can see as reflecting the best strategies for trying to get at the truth—standards which are informed by a heritage of a priori, basic truistic commitments without which inquiry, and meaningful investigation would not be possible. The disclosure of these truistic underpinnings and their rational force is not arrived at by empirical means, akin to a posteriori methods of investigations used in the sciences, but rather by a priori, distinctively philosophical reflection on the preconditions of conceptual grasp and participation in the cooperative pursuit of understanding.

In what follows we shall survey what is wrong with the abductive, and the naturalist, externalist arguments for realism which Smart and Boyd have proposed. Our discussion will be loosely informed by Fine's criticisms, revealing that the best conclusion to be drawn from Fine's insights is not—as he would have us believe—that realists must give up hope, but rather that they ought to favor an internalist, non-naturalist explanationism over an externalist, naturalist one. We shall also see that, in keeping with Fine's wariness about the scope of inference to the best explanation, realists would be well advised to shy away from commitment to certain highly general, unqualified epistemological principles concerning the relation between explanatory superiority and rational compulsion or likelihood of truth (such as "to be rational, an agent must believe the result of an inference to the best explanation", "The explanatory superiority of a theory serves as evidence for its truth", or "explanatoriness is always relevant to likelihood of truth"). The kind of realism which survives these considerations is not the deviant offspring of a fallen program, but entirely true to the aims, ideas, and sentiments that inform any realism that is worthy of the name.

I. SMART'S "WOULDN'T IT BE A MIRACLE?" ARGUMENT

Boyd's philosophical predecessor, J.J.C. Smart, fashioned a fairly simple abductive argument for scientific realism:

What better explanation can there be for the empirical success of a scientific theory T (or, alternatively, the success of its phenomenological counterpart T'), other than the theory's being true? If the theory weren't regarded as true, its astonishing descriptive and predictive capacity would have to be seen as nothing short of miraculous, a great "cosmic coincidence." The most rational thing to do, then, would be to accept that the entities described by science at its best really do exist, and to take every sign of continued empirical success as further indication that our favored scientific theories are true. In Smart's own words:

> I would suggest that the realist could (say) ... that the success of T' is explained by the fact that the original theory T is true of the things that it is ostensibly about; in other words by the fact that there really are electrons or whatever is postulated by the theory T. If there were no such things, and if T were not true in a realist way, would not the success of T' be quite inexplicable? One would have to suppose that there were innumerable lucky accidents about the behaviour mentioned in the observational vocabulary, so that they behaved miraculously *as if* they were brought about by the non-existent things ostensibly talked about in the theoretical vocabulary. (Smart, pp.150f)

This way of putting the argument appears to make it vulnerable to a simple and obvious objection. Smart's claim that the empirical success of a false theory would be a "miracle"—something so outlandish as to defy explanation—seems undermined by the very presence in the history of science of a great number of scientific theories and beliefs which were capable of being empirically powerful for quite awhile, but which we now regard as false. We can, with the benefit of hindsight, tell a story about how these theories, though ultimately false, had the kind of structure which enabled them to map onto the observable part of the world with some success for a while: we can show how, within a certain range of their application, they are isomorphic in what they say about observables to what we now regard as confirmed theory.

So perhaps what Smart really meant to say was that there are some theories whose empirical success is so widespread, long-lasting, and surprising that the truth of such theories has got to be the best explanation for their instrumental reliability. An anti-realist might then try to respond to Smart's explanatory challenge by offering van Fraassen's account of why current scientific theories are so successful empirically: only instrumentally reliable theories survive the test of time and experience because any theory which doesn't is discarded.

> ... I claim that the success of current scientific theories is no miracle.

> It is not even surprising to the scientific (Darwinist) mind. For any sci-
> entific theory is born into a life of fierce competition, a jungle red in
> tooth and claw. Only the successful theories survive—the ones which
> in fact latched on to actual regularities in nature. (van Fraassen 1980,
> p.40)

Alternatively, as Fine notes, the anti-realist might suggest that
a legitimate, alternative explanation for the ongoing instrumental
reliability of T is simply the empirical adequacy of T. The question
"Why is T so remarkably successful empirically?", then, can be
interpreted in a way which invites two sorts of shallow anti-realist
responses: the Darwinian one ("It's no surprise that any current
theory is empirically successful—all the unsuccessful ones get discarded"), and the Scholastic one ("The theory is empirically successful because it is empirically adequate").

Smart might want to object to these anti-realist proposals on
the grounds that they simply involve a rejection of his explanatory
challenge, rather than a genuine response to it. What he really
wants to know is "*How* is the theory capable of being so successful empirically?", i.e., "What features of the theory contribute to
the theory's ongoing empirical success, and why do these features
have this effect?" Appealing to the Darwinian or the Scholastic
explanation here is like answering the question "Why are the
members of this species able to survive, quite surprisingly, in this
harsh environment?" by saying "Well, if they hadn't been fit
enough to survive, they would have died out", or "They have
adapted so well because they have features which are conducive to
survival in this environment." What we really want to know is
which features contribute to the survival of the species, and how
or why they do so.

Even when Smart's explanandum is understood in the proper
way, however, his argument still has problems. For any instrumentally reliable theory T, the anti-realist can in principle propose the
following challenge. Granted, the truth of T is one explanation for
the empirical success of T and its phenomenological counterpart
T', but here is an alternative explanation: the truth of T'', a theory
which is incompatible with T, but empirically equivalent to it.

What features of the former explanation could possibly distinguish it as a better explanation for the relevant explanandum than the suggested alternative? Both explanations do the job they are supposed to do, and they do it in much the same way. There simply aren't any grounds for counting Smart's favored explanation as the best, so his argument fails.

But perhaps Smart would retort by claiming that the alternative explanations in question aren't really available or worked out (even if they could be in principle) because we haven't fully articulated their content. No one, for instance, has bothered to come up with an empirically equivalent, incompatible rival to molecular theory. He might then go on to argue that since realism provides the only completely developed, genuinely available explanation for the widespread, astonishing empirical success of certain scientific theories, and we ought to (or have to) accept the best available explanation for any phenomena that stand in need of explanation, we ought to (or have to) accept realism.

This latest version of the argument won't work either. For one thing, even a realist should reject the premise that we are always rationally constrained to accept the best explanation we can come up with. There are several sorts of situations where sufficient reason for doubt is present, making commitment to the best available explanation optional, if not downright ill-advised. Here are a few commonplace illustrations:

- Fingerprints and DNA traces point to a particular suspect. That the suspect is guilty of the crime in question is the best explanation of the given evidence, but the district attorney feels that strong motivation for a set-up exists, so he won't charge the suspect, or believe that he is guilty until investigators have fully pursued the possibility that the suspect was framed.
- A patient has symptoms which point to the flu as the most likely cause, but, given their duration, a doctor decides to wait until further tests have been performed before making a definite diagnosis.
- Superstring theory accounts for the behavior of subatomic particles, but even if it's the only theory at that level of

causal depth which has been reasonably well-worked out and developed, remaining agnostic about the theory is reasonable: the theory is quite speculative, and the possibility that theorists, given sufficient time and ingenuity, could offer some alternatives seems quite likely.

- Many researchers regard prions as the best available explanation for mad cow disease, especially given the capacity of the deadly agent to withstand assaults (of extreme temperature, radiation, etc.) which ordinary infectious agents like viruses and bacteria cannot survive, but sustaining some skepticism about this hypothesis is fully justified: research is still in its very early stages, and keeping an open mind might become important for making new breakthroughs.

- A scientist in the late 17th century thinks that optical phenomena are in general better explained by assuming light is composed of waves, rather than corpuscles, but also that certain striking problems with Huygens' theory cast some doubt upon it (e.g. the phenomenon of refraction seems to suggest that matter and ether particles interact, yet the failure of material bodies to encounter resistance when passing through empty space seems to suggest that they do not interact).

In general, doubt about or rejection of the best available explanation is reasonable in the following sorts of circumstances:

a. All the evidence we have gathered is compatible with our favored explanation of a given phenomenon, but we admittedly haven't tried very hard to test our hypothesis under conditions where it is most likely to fail.

b. We have made an effort, with limited means, to give as much empirical and theoretical support for the best explanation as we can, but we know there are plenty of decisive tests which we won't be able to perform, or relevant considerations which we are unable to take into account (the subject matter is too complex, or our investigative capacities are too restricted). Pragmatic constraints limit our capacity to pursue alternative explanations which we suspect would turn out to be just as fruitful.

c. The evidence we have which seems to recommend a particular explanation is rather weak; we think that it is entirely possible (if not likely) that undermining evidence will show up later.

    d. A hypothesis is the best explanation relative to our theoretical framework, but the hypothesis fails to secure this status in another equally legitimate research framework.
    e. The best explanation we can come up with still has shortcomings (explanatory loose ends, incompatibility with the results of certain experiments, ad hoc additions...)

We aren't rationally constrained to accept Smart's explanation, even if we were to grant that it counts as the best among the "genuinely available" ones, because sufficiently strong reason for doubt (falling into category b above) is present.

Furthermore, the anti-realist has an even more straightforward way of resisting the conclusion of our latest construal of Smart's "wouldn't it be a miracle?" argument. The argument makes use of inference to the best explanation, a form of inference which the anti-realist regards as illegitimate. We shall return to this sort of objection to the explanationist defense of realism later.

II. BOYD'S ARGUMENTS FOR REALISM

II.i Boyd's Inference to the Best Explanation
According to Boyd's famous abductive argument for realism, realism triumphs as a philosophical position because it provides the best explanation for the overarching empirical success won by the largely theory-dependent methods of scientific inquiry.

Boyd first enjoins us to consider how theory-laden actual scientific practice is. In figuring out how to test a theory, to extend its application, to rate the value of competing explanations, to decide between incompatible theories, to assess the import of data acquired during experiments, to design measurement devises, to choose topics for future research—the many problems and activities that characterize the scientific enterprise—scientists must continuously rely on background theory (experimental protocol, causal principles, theoretical constructs...).

This part of the argument is fairly uncontroversial: evidence of this sort of dependence on background theory is easy to find. For example, in Einstein's Brownian motion defense of the molecular hypothesis, we see how Einstein draws upon background theory to attain a conception of what viable alternative explanations for Brownian motion there could be, what quantitative results would be expected under those alternatives, what quantitative results would be expected under the molecular hypothesis, what types of disturbing factors to control for during experiments...

Take any important piece (and almost any more routine piece) of scientific work, and you will see the influence of immersion in background theory. There is no way to account for what scientists do and what choices they make without acknowledging that they refer back to a rich theoretical heritage.

The next step in the argument is considerably more explosive. Boyd asks a provocative question. What better explanation could there be for the instrumental reliability of the largely theory-dependent methodological principles of science, other than the background theories' (that are relied on in the formation of the principles) being approximately true? Scientific practice, which at every turn depends so heavily on the causally rich heritage of previously confirmed or accepted theory, has been quite effective at producing empirically successful theories. The remarkable success of the methodological principles of this practice is unintelligible unless we grant that we possess approximately true background theoretical knowledge.

Discerning that an important methodological principle in science is good or justified is not an a priori matter. For Boyd, who is a philosophical naturalist, any non-deductive rule of inference is only a posteriori justifiable; it is only as good as the background theory or theories on which it relies, meaning it is only as reliable as its background theories are true. The birth of productive science and productive scientific methods, then, cannot be seen as the natural consequence of applying a priori principles of rationality to experience, but rather as a historically contingent outcome made possible by the rather lucky accident of happening upon an approximately true theory or theoretical framework (which could then be drawn upon for future research).

Judgments of explanatoriness in science (a variety of what Boyd calls "judgments of theoretical plausibility") are relevant to

judgments about likelihood of truth because our ratings of explanatory superiority are determined by causal principles and theoretical
considerations which are a part of our heritage of approximately
true theoretical knowledge. Principles for evaluation informed by
background scientific theory indicate that a certain hypothesis
among competing alternatives is most plausible and provides the
best explanation for a given phenomenon. This state of affairs
speaks in favor of the approximate truth of the hypothesis.

> Central to the realists argument is the observation that projectibility
> judgments are, in fact, judgments of theoretical plausibility: we treat
> as projectible those proposals that relevantly resemble our existing
> theories (where the determination of the relevant respects of resem-
> blance is itself a theoretical issue). The reliability of this conservative
> preference is explained by the approximate truth of existing theories,
> and one consequence of this explanation is that judgments of theo-
> retical plausibility are evidential. The fact that a proposed theory is
> plausible in light of previously confirmed theories is some evidence
> for its (approximate) truth. Judgments of theoretical plausibility are
> matters of inductive inference from (partly) theoretical premises to
> theoretical conclusions; precisely these inferences justify, and explain
> the reliability of, 'inductive inference to the best explanation'. (Boyd
> 1996, p.224)

The best explanation (as rated in accordance with previously
confirmed theory) is more likely to be true than alternatives which
are explanatorily inferior (again, as assessed in light of previously
confirmed theory)—even if these alternatives are empirically equivalent to the best explanation. Why? Because, according to Boyd,
our background theories are approximately true. And why must
we think that our background theories are approximately true?
Because, if we didn't, we wouldn't be able to account for the amazing instrumental success of the heavily theory-dependent methods
of science.

Here we have, then, an attempt to defend what are often
regarded as two central realist tenets:

1. We have approximately true theoretical knowledge (we
   have to accept that some of the unobservable entities
   posited by our best scientific theories exist).

2. Judgments of explanatory superiority are relevant to judg-
    ments of likelihood of truth (the explanatory superiority of
    a theory counts in favor of its truth).

A problem seems to emerge, however: Boyd's defense of 1
relies on 2 (Boyd uses inference to the best explanation to argue for
1), yet the strength of 2, as assessed in Boyd's program, seems to
rely on 1.

The circularity I've identified here is certainly related to the
one first highlighted by Arthur Fine:

I. In his defense of realism, Boyd invokes inference to the best
explanation, but isn't the epistemic status of explanatoriness
exactly what is really at issue in the debate between realists and
anti-realists (the realists accepting and the anti-realists denying
that explanatoriness is relevant to truth)? By relying on abduction
to establish the truth of realism, Boyd is simply begging the question against anti-realism.

Though I think that this indictment is, among Fine's barrage of
criticisms of Boyd's explanationist defense of realism, the one which
is most helpful in uncovering what is truly amiss in Boyd's philosophical program, other objections Fine raises are worthy of note:

II. In general, anti-realists tend to reject the idea that genuine
needs for explanation exist in the first place, so why would they have
to accept Boyd's claim that the instrumental reliability of the theory-
dependent methods of science stands in need of explanation?

III. Even if we ignore the circularity problem, we can still
worry that Boyd neglects to consider alternative anti-realist explanations for the success of theory-dependent methods.

IV. Furthermore, it's not so clear that Boyd's explanandum
holds. The history of science does not present a series of increasing
triumphs: it is marked by error and defeat. The theory-dependent
methods of science have led to successes, but also a great many
failures, so do we really have good reason for thinking that our
background theories must be approximately true?

I shall discuss Boyd's attempts to respond to some of these criticisms, though I shall concentrate on his tellingly unsuccessful
efforts to dampen the impact of I.

II.ii Rival Explanans and Explanandum

Let's first look at IV. In order to assess how well Boyd handles this sort of objection, we need to have a better understanding of how and why evidence of repeated falsehood and failure in scientific inquiry potentially threatens Boyd's project.

We saw how, under a certain reading, Smart's defense of realism seems to be called into question by the presence of scientific theories which have enjoyed some measure of empirical success, but have ultimately turned out to be false. Productive failure in the history of science shows that the best explanation for the instrumental reliability of a theory cannot always be the truth of the theory. But perhaps we can raise a similar problem for Boyd: these sorts of fruitful mistakes also seem to indicate that the best explanation for the instrumental reliability of theory dependent methods cannot always be the truth of the background theories from which they are derived; sometimes these previously accepted theories turn out to be false. If this sort of success cannot always be explained by the truth of background theory, why should it ever have to be?

This objection is not quite the one Fine ends up with in his exploration of the implications of scientific failure for explanationist defenses of realism. Fine does not say that our experience of the possibility of instrumentally reliable methodological principles which are grounded in false background theory casts some doubt on the need for or plausibility of a realist explanation of the empirical success of scientific practice in general. Instead, he emphasizes how often the theory dependent methods of science lead to error and falsehood. We don't need to explain why scientific practice succeeds so brilliantly—it doesn't. If anything, we need to explain why methods, which so often lead us astray, sometimes triumph. It's far from clear that a realist's explanation for this pattern of failure dotted by occasional success would be especially satisfying.

> ... in formulating the question as to how to explain why the methods of science lead to instrumental success, the realist has seriously misstated the explanandum. Overwhelmingly, the results of the conscientious pursuit of scientific inquiry are failures: failed theories,

failed hypothesis, failed conjectures, inaccurate measurements, incorrect estimations of parameters, fallacious causal inferences, and so forth. If explanations are appropriate here, then what requires explaining is why the very same methods produce an overwhelming background of failures, and, occasionally, also a pattern of successes. (Fine 1996, p.27)

In "Realism, Approximate Truth, and Philosophical Method", Boyd hopes to present a legitimate account of approximate truth which will be of some help in overturning the objection that evidence of productive failure undermines the appeal of his realist explanation for the instrumental success of the methods of science. Though he does not directly address Fine's worry that realists have grossly overestimated the level of success these methods have attained, Boyd would no doubt anticipate that his notion of approximate truth would go some way towards easing this concern as well.

Boyd concedes that we shall never be able to articulate an abstract, context-independent theory of approximate truth on the model of Tarski's theory of truth; a highly general explication of the notion of approximate truth is unattainable. To make a proper judgment concerning whether a theory should be regarded as approximately true, or simply false, requires sensitivity to context-dependent considerations: an awareness of historical setting, what interests have governed the promotion of the theory, what features are to be taken to be most central, what type of questions the theory was originally designed to address, what types of phenomena it can still accommodate, what aspects of the theory remain fruitful in the subsequent development of science...

A theory which is right about some things, wrong about others, or correct in certain respects, wrong in others can be legitimately regarded as approximately true (instead of simply false) when, for example, the theory productively addresses the sort of major problems it was designed to solve, but also contains less central, potentially dispensable parts which are false. What counts as a less essential part (and whether the falsehood of certain parts does or doesn't undermine the idea that the theory is approximately true) cannot be decided in a context-independent way: there are no formal rules or episode-neutral specifications.

A theory can be the source of effective methodological principles, and play a vital role in the development of new scientific insights, even if it is only true in certain relevant respects. When a realist tries to show how the success of methods at a certain stage in scientific inquiry is best explained by the approximate truth of background theory, she must demonstrate how background theory, though perhaps strictly speaking false, can be regarded as true in certain relevant respects, respects which help explain why methodological principles drawn from the theory could turn out to be so effective. (Of course, to preserve the idea that the explanation is genuinely realist, the relevant respects in which the theory turn out to be true can't simply be "truth in what the theory says about certain observables"!) For example, she might try to highlight how the approximate truth of Newtonian mechanics, the truth of certain relevant parts of Newton's theory (the identification of the principle of inertia, as well as quantitative laws governing the motion of non-subatomic particles with velocities much lower than the speed of light) helps explain how the theory could contribute to the formation of productive methodological principles and play a role in the development of future physics, even though some other parts of the theory are false.

Realists of Boyd's stripe, then, have some resources for handling the objection that the history of productive failure in science alone shows that the triumphs of theory dependent methodological principles don't require a realist explanation. Boyd would insist that, in cases of productive failure, the best explanation for the instrumental reliability of the resulting methodological principles would have to be at least the approximate truth of strictly false background theories.

Boyd accepts that some philosophers might be quite wary of the absence of a topic and episode-neutral standard for the assessment of approximate truth. With the license to make use of theoretical and context-specific considerations, won't a realist always, when faced with a productive, false background theory, be able—

in a contrived sort of way—to point out some respects in which the background theory is or could be true, and thereby save the realist explanation of the instrumental reliability of the methods derived from it?

To allay this fear, Boyd insists that the so-called "contrivance" here is no more illegitimate than the application of Darwin's theory of evolution to explain the presence of certain features of organisms. There is no context or theory-independent way of assessing what evolutionary mechanisms might be at work in a particular case, but that doesn't make the explanation we finally end up with ad hoc or unsavory.

The comparison with piecemeal Darwinian explanation is partly helpful in putting anti-realist qualms about contrivance to rest. It seems to allow Boyd to argue that the absence of a highly general method for determining which features of a theory should count as approximately true does not, by itself, show that his realist interpretation of the history of science is recklessly self-fulfilling. The Darwinian parallel allows Boyd to demonstrate that realism is not obviously undermined by the history of frequent failure in science. Nonetheless, the anti-realist can still object that Boyd's piecemeal appeals to approximate truth in his explanations for the success of theory dependent methods diverge from Darwinian explanations in a significant and revealing way: the phenomena in question have competing anti-realist explanations (which appeal, for example, to the empirical adequacy of the theories in question instead of their approximate truth), whereas in the Darwinian cases there are no satisfactory non-Darwinian explanations.

Boyd's account of approximate truth might be partially helpful in addressing Fine's "failure in science" objection as well. Though the realist is most interested in demonstrating how and why the approximate truth of background theory best explains the instrumental reliability of the theory dependent methods of science, she can also try to show how, for certain episodes in the history of science, falsehood in the relevant respects in background theory have led to the development of unsuccessful methods. Whether you feel, as Fine does, that the failures stand out more than the successes, or, as Boyd seems to, that the successes outshine the failures, the realist seems to be capable of offering a coherent realist explanation: highlighting failure in science alone will not suffice to show that the realist's explanation is unavailable or inadequate.

Whether the realist explanation (under either construal of the explanandum) is genuinely the best, the only one, or one we should accept has not yet been settled. Returning to objection III, we can wonder whether explanationists like Boyd give enough credit to the sort of explanation for the success of theory dependent methods which anti-realists have to offer, namely that our theories are more or less empirically adequate. That we end up possessing empirically adequate theories is no surprise, for the ones which aren't get weeded out (just as it's no surprise that the species of the world have features which are well suited for survival in their habitats: the species that don't have such features become extinct). An anti-realist might also try to explain why our scientific methods are so successful by suggesting that the theories depended upon in the development of these methods are perhaps largely false in what they say about unobservables, but nonetheless empirically equivalent to what would be true theories. These explanations become available if you believe, as anti-realists do, that the success of methods based on confirmed or accepted theories doesn't speak against the possibility of alternative theories (which are empirically equivalent to, but incompatible with our accepted theories) serving as the source of equally successful methodological principles.

If pointing to the empirical adequacy of scientific theories (or their empirical equivalence to approximately true theories we don't possess) to account for the instrumental reliability of scientific methods seems too much like a scholastic appeal to dormative virtues—an evasion of Boyd's explanatory challenge, rather than a response to it—we can ask whether the demand for explanation is one which must be met. If our scientific theories are not even approximately true in a sense that is relevant to the debate at hand, "really" showing how they could nonetheless be the source of instrumentally reliable methods would require having access to a

God's eye view. We would have to be able to demonstrate, for instance, how certain features of our theories bear such and such a relation to the structure of reality, such that it becomes clear how working with our models of the world leads to the generation of instrumentally effective causal principles and investigative procedures, even though our theories fall far short of capturing the true structure of the world.

Conceding that our epistemic limitations prevent us from being able to meet the demand in the deep way realists might like doesn't mean that we have to accept the only fully worked out alternative (in this case, Boyd's realist explanation). Neither does this concession show that we are committed to saying that the phenomena in question don't need to be seen as having an explanation (even in principle or in fact). I mention this point in order to emphasize that the kind of a threat reflected in objection II is quite distinct from the one raised in III. In a Humean mood, an anti-realist can be radically suspicious of the idea that any regularity or phenomenon stands in need of explanation, or must be seen as having some explanation or other (however inaccessible); this is the intuition that drives II. On the other hand, an anti-realist might sometimes feel that, in the interest of securing empirical adequacy, a given theory ought to be able to "save the phenomena", and such efforts to save the phenomena could perhaps be legitimately regarded as attempts to explain the phenomena (though of course the fact that one of two conflicting, but equally empirically adequate theories provides a "better" explanation than the other does not, for the anti-realist, speak in favor of it's truth, or likelihood relative to the explanatorily inferior theory).

Whether a philosophical theory ever falls into this category is not so clear (not everyone is a philosophical naturalist like Boyd); but, to be on the safe side, an anti-realist might try to work with Boyd's assumption that a defensible philosophy of science ought to be able to save the relevant phenomenon, the success of scientific methods, and show that anti-realism is at least as adequate empirically as realism. That the resulting explanations might have to be rather shallow (and perhaps even faintly scholastic) because the causally deeper explanations are epistemically inaccessible would hardly bother the anti-realist: for her, empirical adequacy is relevant to truth, explanatory depth is not.

So how would Boyd himself regard the suggestion that anti-realists can reject his explanation in part because they can provide, by their own lights, reasonable—albeit shallow—anti-realist explanations for the instrumental reliability of the theory-dependent methodological principles of science?

One gathers he would have little patience for such a response. He seems quite convinced that the anti-realists simply cannot develop plausible explanations for any number of important things, such as the possibility of purely instrumental knowledge or the continuity of reference despite revisions in definition, let alone the individual triumphs (the reliability of measurement and computational procedures, of features of experimental design, of projectibility judgments ...) that together constitute the phenomenon in question. At best, the anti-realist must resort to her blasé attitude about the need for explanation in the first place.

> The key argument for scientific realism according to the programme presented here is that realism as a scientific hypothesis presents the only scientifically acceptable explanation for the reliability of scientific methods. The empiricist might be unimpressed by the demand for explanation in this case ... Still, the realist can also argue that accepting the realist explanation provides as well the only justification we have for accepting the instrumental findings of science. (Boyd 1996, p.251)

What partly motivates Boyd's confidence that any anti-realist explanation is unacceptable or inferior is his conviction that the empiricist "evidential indistinguishability" thesis (empirical evidence should affect our judgments of the likelihood of empirically equivalent theories equally) is flawed and misguided. Raw data should not affect our judgment of the likelihood of empirically equivalent theories equally, for theoretical plausibility in light of an entrenched, successful research tradition is itself, according to Boyd, "evidential": an additional factor to be taken into account when making judgments about likelihood. An alternative which is judged to be most plausible relative to previously confirmed theories should be accorded a greater likelihood than some empirically equivalent, contrasting alternative. In other words, we should adhere to a norm of methodological conservatism in deciding which among empirically equivalent, incompatible alternatives is most likely to be true.

But what is the justification for methodological conservatism? Why should we accept that theoretical plausibility is evidential? I'm not quite sure how Boyd would answer this question, but two types of responses seem present in his writings:

1. Methodological conservatism, as a norm which is part of scientific practice, has shown itself to pay off well in the continued instrumental success of scientific methods. Violating methodological conservatism (by pursuing a radical, empirically equivalent alternative to a theoretically plausible hypothesis, a radical alternative which would be most plausible only relative to some non-actual, total science which is empirically equivalent to our set of current, confirmed theories) is not a part of ongoing scientific practice, so we have no reason to think that the overarching results would be as favorable. The burden of proof is on the anti-realist to show that they could be as favorable.

2. The approximate truth of our background theories accounts for the effectiveness and appropriateness of our conservative preference for hypotheses which are most plausible in light of background theory (see the quote from "Realism, Approximate Truth, and Method" above). And why should we accept that our background theories are approximately true? Because the success of the methodological principles of science (which include methodological conservatism, and its offshoot, inductive inference to the best explanation) are best explained on this assumption.

Neither of these responses strike me as satisfactory, for the anti-realist is free to argue in the following way:

Against 1: To violate methodological conservatism is to resist using your traditional theoretical framework to make choices between equally empirically adequate hypotheses or theories. Such a violation may involve embracing a complicated, radical, and somewhat contrived alternative to a widely accepted view (of the sort Poincaré discusses). For example, a non-conservative scientist could abandon the molecular hypothesis, and choose to try to reinterpret all the phenomena the molecular hypothesis explains by developing an elaborate new theory of the continuity of matter—a theory which posits all sorts of novel entities or fields, and elaborate quantitative laws which secure empirical adequacy at the cost of elegance and simplicity.

Violating methodological conservatism may also be achieved somewhat less spectacularly by abandoning your original theoretical framework, in favor of an alternative existing framework which is just as adequate empirically, but no more defensible in a framework-neutral way than your old one. For example, a scientist of the 17th century could have converted from the Newtonian corpuscular theory of light to Huygens' wave theory, without necessarily being motivated by epistemically relevant considerations. (We assume, for the purposes of this example, that with enough ad hoc maneuvering, the corpuscular and wave theory could have been regarded as equally adequate empirically in the relevant time frame.)

Of course, in practice we rarely depart from methodological conservatism, but the fact that such a departure from theoretical tradition is difficult to motivate is not a sign of the epistemic irrationality of violating methodological conservatism. These violations have little to recommend for themselves because they are unsatisfactory and cumbersome from a practical standpoint. Their viability epistemically should not necessarily drive us to consider adopting them, but it should have an influence on our probability assignments (no theory should be assigned a value higher than any other equally empirically adequate alternative).

We have no reason to think that the overarching results of violating methodological conservatism would not be as favorable as following the norm. If anything, we have a reason to think that the overarching results *would* be as favorable: given the empirical equivalence of the radical, or unfamiliar tradition with the familiar tradition, and the empirical equivalence of the hypotheses most plausible relative to the former and the latter, how could a violation of methodological conservatism make a difference empirically? Contrary to what Boyd thinks, the burden of proof is on him to show that it would.

We have, then, a clash of intuitions between Boyd and the anti-realist. Boyd believes that the burden of proof is on the anti-realist to show that violating methodological conservatism could be as successful as following it, whereas the anti-realist thinks that the burden of proof is on Boyd to show that violating methodological conservatism could make a difference empirically. To gain any ground at this point, and to resist the charge of dogmatism in his argument for realism, Boyd would need to make a case (which is fair to all parties involved) for thinking that his conception of the burden of proof is the best one.

Against 2: Boyd's line of reasoning, which itself makes use of inference to the best explanation, begs the question against anti-realism: the anti-realist rejects Boyd's assumption that the explanatory superiority of a theory speaks in favor of the theory's truth.

We have finally made our way to the circularity objection.

II.iii The Circularity Objection and the Realist Package

How can Boyd himself rely on inference to the best explanation in his primary defense of realism, when the epistemological status of inference to the best explanation is precisely what is at issue in the debate between realists and anti-realists?

One route Boyd could take would be to draw on resources available within his program to try to show that inference to the best explanation is a legitimate form of inductive reasoning which leads us to accept theories which are likely to be true, or more likely to be true than their explanatorily inferior rivals, even when these rivals are empirically equivalent to the best explanation. In other words, he could try to defend the idea that judgments of explanatoriness are evidential.

Boyd would, presumably, have to model such a defense on his justification for the claim that judgments of theoretical plausibility are evidential: judgments of explanatory superiority are, after all, essentially based on, if not identical to, judgments of theoretical plausibility.

1. Generally accepting the results of an inference to the best explanation, as a norm which is part of scientific practice, has shown itself to pay off well in the continued instrumental success of scientific methods. Violating inference to the best explanation (by pursuing a radical, empirically equivalent alternative to the best explanation, a radical alternative which would count as the best explanation only relative to some non-actual, total science which is empirically equivalent to our set of current, confirmed theories) is not a part of ongoing scientific practice, so we have no reason to think that the overarching results would be as favorable.

2. The approximate truth of our background theories accounts for the effectiveness and appropriateness of our preference for hypotheses which are the best explanation, relative to background theory . And why should we accept that our background theories are approximately true? Because the success of the methodological principles of science are best explained on this assumption.

Naturally, any reservations one might have had about the arguments for theoretical plausibility being evidential will emerge again here. The anti-realist will insist that the fact that inference to the best explanation is part of an empirically successful, entrenched practice gives us no reason for thinking that violating inference to the best explanation by making empirically equivalent choices wouldn't also have been as successful. She will also, given her deep reservations about the rational force of inference to the best explanation, find the appeal to inference to the best explanation as a step in securing its appropriateness as a form of inference viciously circular. When the relevant opponent is the anti-realist, and what is at stake is the view that belief in certain unobervables is not simply rationally acceptable, but rationally compelled, these efforts at justification are inadequate.

Perhaps Boyd wouldn't, in the end, feel drawn to these two unsuccessful defenses. After all, it has not entirely escaped his attention that his use of abduction to defend realism has aroused some suspicions. In "Realism, Approximate Truth, and

Philosophical Method", he openly admits that his abductive argument for realism works only if the realist explanation is interpreted or understood "realistically." To dispel the illusion that he is simply preaching to the converted, he concedes that the abductive argument should not be seen as capable of standing on its own. The realist explanation for the success of the methodological principles of science is only one part of a whole set of philosophical contributions and solutions that realists have to offer, a "philosophical package" which allows us to address pressing issues in epistemology, metaphysics, and philosophy of language which even an anti-realist would want to take seriously. Armed with a naturalist realism, Boyd demonstrates how, for example, we can:

- secure metaphysically robust, non-Humean accounts of causal relations
- use reliabilist accounts of knowledge to avoid the perplexities and problems inherent in foundationalism (e.g. by developing a non-foundationalist treatment of projectibility judgments)
- introduce naturalized, causal accounts of reference to account for the patterns of apparent essentialism that we seem to see in scientific language.

In two earlier papers, "On the Current Status of Scientific Realism" and "Lex Orandi est Lex Credendi," Boyd elaborates on some of the details of this sort of strategy for undermining the circularity objection, and tries to show that the capacity of realism to account for the instrumental reliability of the theory dependent methods of science is only one of the superior features of his naturalist, reliabilist program. He argues that his naturalist realism can accommodate or motivate certain practices which all but the most radical anti-realists would want to regard as fully legitimate or rationally required, practices which, within anti-realist epistemology, problematically turn out to lack justification.

According to Boyd, the anti-realist cannot take the following practices as well-motivated without accepting the sort of theoretical commitments realism requires, and thereby capitulating to the realists:

1. Following the principle of the unity of science; the anti-realist cannot account for why integrating independent, empirically successful theories is often successful.

2. Forming assessments of empirical adequacy or instrumental reliability (assessments which are, essentially, projectibility judgments); although most anti-realists do not wish to undermine our judgments about the empirical adequacy of theories, or the instrumental reliability of certain methods, they don't have the resources to defend the idea that such judgments are justified.

Here is how Boyd tries to defend these claims.

1. Successful unification of theories which we witness in science only makes sense under the realist assumption that the theories which are integrated present a fairly accurate description of the way the world is. Taken independently, the theories identify and describe the relevant causal mechanisms in ways which may not always link up to empirical evidence directly, but do allow for extension or incorporation into a new, broader theoretical framework once the appropriate adjustments have been made. (Unification, after all, is much messier than simply "taking the conjunction" of two different theories.) Given the considerable amount of theoretically motivated tampering which accompanies unification, if we were to regard the independent theories as merely empirically adequate, we would have no reason to anticipate that they could be successfully combined, and no way of accounting for the fact that unification is often effective.

2. Goodman's new riddle of induction illustrates that we need to appeal to background theory to single out certain predicates as genuinely "projectible" among the infinitely many inductive generalizations which are, strictly speaking, compatible with any given set of instances—even where these instances are said to speak in favor of the empirical success of a scientific theory. The claim that a scientific theory is empirically adequate (or that a particular scientific method is instrumentally reliable) is intended to indicate confidence in past, present, and, to some extent, future compatibility with the relevant observable facts. Theory dependent projectibility judgments are required in order to make appropriate

inductive generalizations about what the relevant observable facts have been and will be, and which theories accommodate these facts.

Without the background theoretical commitment, choosing a particular inductive generalization over one of the infinitely many alternatives which are compatible with the data becomes arbitrary and unjustified. Since the anti-realist, unlike the realist, refrains from adopting theoretical commitments, her acceptance of the inductive generalizations (even about observables) that are involved in making judgments about empirical adequacy or instrumental reliability are, from the anti-realist's perspective, unjustified.

That anti-realists routinely need to rely on, or want to rely on inferences which have no justification within anti-realist epistemology seems to be an even greater problem than the fact that some realists will try to support realism by means of a kind of inference (namely, abduction) which is justified only within realism. Circularity isn't ultimately as bad as engaging in practices at odds with what you preach. We should also keep in mind that the defense of realism is a broader philosophical project than might be suggested if we focus on the traditional, circular explanationist strategy.

Boyd wants us to tally up the individual successes and failures when we compare how well anti-realism and realism (that is, reliabilist, naturalist realism) fare at answering questions we should all want an epistemological theory to answer. The failure of anti-realism to legitimize even ordinary claims about empirical adequacy is a significant shortcoming.

Problems with 1

2 seems to present a newer and more serious challenge to anti-realism than 1. 1 simply highlights yet another methodological principle, the success of which, according to Boyd, the anti-realist cannot explain, so 1 may be weakened by the sorts of objections to Boyd's abductive argument for realism which we have already examined:

I. Boyd is implicitly relying on abduction when he infers that the anti-realist's inability to account for the instrumental reliability of the principle of the unity of science makes anti-realism less plausible, so he is begging the question against anti-realism.

II. The anti-realist resists the idea that explanations for phenomena are ever genuinely required, or that certain types of phenomena genuinely stand in need of explanation in the first place.

III. The anti-realist can account for the empirical success of unification: unified theories tend to be empirically successful, because the ones that aren't fail to survive.

IV. Unification has led to failure at least as often as it has lead to success, so there is no general instrumental reliability which we need to account for.

III is an important part of a response by the anti-realist who thinks that Boyd's challenge is worth taking seriously. Some anti-realists will be convinced that unification is a non-arbitrary, well-motivated scientific practice, and that it is part of the job of a philosopher of science to determine what makes scientific methodology rational and justified, and what sometimes prevents non-scientific or pseudo-scientific methods from counting as rational. This sort of anti-realist will be unsettled by the circularity of Boyd's appeal to abduction, but will still feel the need to illustrate why individual scientific practices, like unification, are well-motivated.

Unlike Boyd, the anti-realist feels that any norm of scientific inquiry will be justified to the extent that following that norm can be seen to contribute to the development of useful, empirically adequate theories. Unification is a well-motivated scientific practice, when it does not compromise empirical adequacy, because it makes theoretical machinery less cumbersome and complicated. The quest for unification has been a somewhat empirically successful venture in science because unifications which fail to be empirically successful are thrown out.

The anti-realist can cast further doubt on the need for Boyd's realist understanding of why scientists follow the principle of the unity of science by asking: why should we expect that merging two independent theories, which we take to be approximately true, will result in a unified theory which will also be approximately true—

any more so than we would expect that merging two independent theories, which we take to be empirically adequate, will result in a unified theory which will be empirically adequate? Merging only widens the error bar, perhaps so much so that it would be an abuse of language to say that the resulting theory is approximately true.

Problems with 2

Boyd regards claims about empirical adequacy as projectibility judgments—a reasonable assessment—but there are at least two quick responses an anti-realist might want to make to the charge that the anti-realist must either embrace the relevant theoretical considerations and accept defeat on this issue, or reluctantly accept that these routine projectibility judgments are unjustified:

i. In keeping with his view that theoretical commitments in science are rationally acceptable, but never rationally compelled, van Fraassen might wish to address the problem Boyd has raised by accepting that the theoretical commitments involved in making judgments about empirical adequacy are rationally acceptable, but not rationally required. Accepting alternative projectibility judgments about instrumental reliability which conflict with our own is not epistemically irrational, though it may be irrational from a practical standpoint.

ii. A more radical anti-realist, like Feyerabend, may feel that our projectibility judgments (including generalizations about empirical adequacy) are justified only in a very weak sense: there is a rationale for such judgments relative to our background theoretical framework, but the framework we actually use is, epistemically, no better than alternative frameworks, and our choice is simply a matter of sticking to tradition. Given the arbitrary nature of our preference, there is a sense in which our projectibility judgments are unjustified: there is no framework-neutral way of defending our projectibility choices against proposals which alternate frameworks suggest.

A third, more weighty objection to Boyd's argument against anti-realism should be appealing to anti-realists and non-naturalized realists alike.

iii. Boyd's challenge to the anti-realist to try to justify our ordinary claims about instrumental reliability (without making realist assumptions) can be recast as a version of Hume's problem: how do we justify our inferences from instances of a scientific theory's successful accommodation of certain types of observable phenomena in the past, to the generalization that the scientific theory will continue to (or in general does) accommodate these kinds of observable facts? The sort of inference involved is simple induction, but if we can only draw upon experience to legitimate claims we make about the world, or to secure that the methods we employ to find out about the world are good ones, then how can we begin to go about establishing that our basic, non-deductive methods for inquiry into world (including ones of the sort we need for making generalizations about empirical adequacy) are in order? We would, presumably, need to use non-deductive inference rules to learn anything from experience beyond the content of our immediate experiences, so we wouldn't be able to begin to appeal to experience to justify these basic methods of inquiry without first presupposing them. The only sort of justification available for basic non-deductive methods is a viciously circular one, and hence is no kind of justification at all.

Boyd has, apparently, an easy way out of this notorious problem: a non-deductive form of inference is justified if it is, in fact, a reliable procedure for getting at truths (even if the agent has no idea that the procedure is in fact reliable, and even if the agent could never, in principle, conclusively establish that the procedure is reliable). The projectibility judgments which underlie our generalizations about the empirical adequacy of scientific theories (or the instrumental reliability of particular scientific methods) are justified because they rely on background theoretical considerations and corresponding basic methods of inference which are in fact true and reliable.

Does the reliabilist analysis allow us to solve Hume's problem once and for all, or is Boyd ultimately in at least as much trouble as his anti-realist opponents when it comes to trying to address the problem without appeal to vicious circularity, infinite regress, or dogmatism? The reliabilist wants us to interpret the usual terms of epistemic appraisal ("rational," "reasonable," "justified," "known") in a strictly externalist fashion, but we can certainly wonder whether something significant gets lost in the translation.

Even though the externalist accepts that a belief or method's being rational or justified depends, ultimately, on the obtaining of circumstances cognitively inaccessible to the agent, she will have to concede that agents are capable of making good judgments about when beliefs and methods of inference should count as rational or justified, and realizing that their judgments on these matters are reasonable and good; otherwise, the terms for epistemic evaluation like "justified" and "rational" could have no place in our language: we would never know when we could responsibly apply them.

In line with this concession of the part of the externalist, Boyd accepts that legitimate non-deductive rules are not only justified (i.e. "reliable methods for getting at truths"), but also that such rules have or can be given a justification, where "a justification" is taken here to mean "an account of why relying on a method of inference is a good, reasonable thing to do, given our interest in following reliable procedures for getting at truths." As a naturalist, he is committed to the view that these justifications must appeal to empirical considerations alone:

> "If the distinctively realist account of knowledge is sound, then the most basic principles of inductive inference lack any a priori justification."(Boyd 1983, p.211)

> "...there are no a priori justifiable rules of non-deductive inference, and it is an a posteriori question about any such inference whether or not it is justifiable."(Boyd 1990, p.227)

The only sort of account of why relying on the basic principles of inference is a good, reasonable thing to do (given our interest in following reliable procedures for getting at truths) is, according to Boyd, an empirical or "scientific" one, but this naturalist claim invites a version of Hume's problem all over again. If the defense of the idea that it is good for truth-seeking agents to follow the most basic principles of non-deductive inference must be empirical, what non-deductive inference rules can be appealed to in order to learn, from experience, that the basic non-deductive inference rules are good for truth-seeking agents to follow? If we say "the very same rules we are justifying", we are caught up in a vicious circle; if we say "even more basic non-deductive inference rules", we are taking the first step in an infinite regress; if we say "the basic non-deductive inference rules are justified in light of experience simply because they are in fact reliable", we are simply being dogmatic (not to mention the fact that we are equivocating on what giving a justification in this context means). The externalist "solution" to Hume's problem which Boyd offers is really no solution at all.

III. NATURALIZED VS. NON-NATURALIZED REALISM

A non-naturalized, internalist realist does not insist that the only sort of defense of the appropriateness of our acceptance of our most basic beliefs and methods is wholly empirical, so she has more options when in comes to addressing Hume's problem than the unsatisfactory ones we have just rehearsed. She can take the transcendental route, and urge us to see the legitimacy of relying on these basic beliefs and methods as grounded in their acceptance being a precondition for the very possibility of meaningful inquiry, or cognitive grasp of the world. She can say, along with Strawson, for example, that we do not come to think that we ought to follow the "rules" of basic induction because experience tells us that basic induction generally works well (Hume has adequately demonstrated the futility of this response); instead, the legitimacy of the rules of basic induction is grounded in the fact that anyone who wants to be doing something that can legitimately be called "empirically investigating the world" must follow these rules. If the uniformities in nature were to suddenly change, that wouldn't show that our reliance on basic induction was irrational or ill-advised. In such a circumstance, we wouldn't abandon basic induction; we would need to appeal to it to form appropriate generalizations from experience about how various phenomena, which appeared to display a certain regularity, no longer do so.

These conclusions show, I think, that it really isn't so easy to put realism, naturalism, and reliabilism together after all, despite Boyd's feeling that the best sort of philosophy of science includes them all. Hume's problem is a challenge for any package, but Boyd's seems particularly fragile. Boyd's efforts to show that charges of circularity are counterbalanced by his realism's success at solving this old epistemological puzzle appear to have backfired: the traditional problem of induction seems to expose more weaknesses than strengths in his naturalist, externalist project—weaknesses which are not present in non-naturalized realism.

Our examination of the unity of science principle and the problem of induction suggests that Boyd's realism doesn't really provide satisfactory solutions to more problems. Furthermore, we can cast doubt upon whether it's such a good idea for a realist to be a philosophical naturalist (let alone whether a philosophical naturalist should try to defend realism by sizing up the relevant packages) in the first place.

Since Boyd is a philosophical naturalist, he sees philosophy as essentially continuous with empirical science. To the philosophical naturalist, there is nothing truly distinctive about the methods of philosophical inquiry. There are no special tools to which philosophers have access for securing new discoveries or justifying theories: philosophical views must be defended in much the same way scientific views are (without, of course, the heavy stress on elaborate experimentation and quantitative results). It should not come as a surprise, then, that Boyd would see the rational force of realism as grounded in the comprehensiveness of realism (it's ability to provide solutions in several areas in philosophy), for scientific theories can achieve their status for much the same reason: their ability to unify disparate phenomena, to solve a number of different problems at once.

Boyd's commitment to philosophical naturalism appears to open up new avenues for the defense of realism which avoid the perils of reliance on abduction, but we are still left to wonder whether such a commitment burdens Boyd with a threat of inevitable, irredeemable circularity—a circularity from which there is no escape.

If the methods we must use for justifying philosophical theories cannot depart from the methods we use for justifying scientific theories (as the philosophical naturalist claims), and whether the methods of science ever lead us to theoretical beliefs which must be seen as approximately true or rationally preferable among all alternatives (from an epistemic point of view) is precisely what is at issue in the debate between realists and anti-realists, then isn't it question begging to use these methods to show that realism is rationally compelled or approximately true? The anti-realist denies that the results of the application of scientific methodological principles ever lead to theoretical commitments which must be seen as rationally preferable to all alternatives or approximately true, so why would she have to regard the results of their application in philosophy of science (here, theoretical commitment to realism) as rationally compelled or approximately true?

Given that the naturalist, by self-avowal, can only use methodological principles which are scientific, the realist who stays true to philosophical naturalism will never be able to defend realism in a non-circular fashion.

So should realists relinquish all hope of adequately defending realism?

Fine would no doubt want us simply to give up and adopt his "natural ontological attitude", a view which retains something akin to Boyd's naturalism, but allegedly transcends the imperfections of both realism and anti-realism, not least by being casual and sensible.

The "NOA-er" (as Fine would put it) believes that philosophy has very little to contribute to the task of interpreting the results of science, or understanding its multifarious aims. Drawing a conclusion that would no doubt please the many physicists who feel suspicious of philosophers who poke their noses in other people's fields. Fine insists that the best place to look for answers to these questions is within science itself. That an active debate about realism in philosophy of science persists is symptomatic of widespread misunderstanding.

According to Fine, realists and anti-realists alike suffer from the misapprehension that they have special tools which enable them to discern the real goal of scientific investigation (developing theories that are true about the world; developing theories that are empirically adequate). Their adherence to inflexible, yet inadequately defended epistemological principles (that we must believe/never have to believe the result of an inference to the best explanation; that explanatoriness is always/is never relevant to truth) makes them think that they are in a unique position to decide what sort of cognitive stance should be taken toward the confirmed results of scientific inquiry (belief in the existence of particular unobservables; mere acceptance of unobservables). Contrary to what the realists and anti-realists believe, no one in fact has better access to the considerations that fruitfully settle this issue than the working scientist.

Abstract philosophical reflection from above is idle and irrelevant. When deciding whether to believe a theoretical claim, or merely accept it as empirically adequate, we should trust the best judgment of the practitioners of science. What type of response is most appropriate will then vary from case to case, for science taken "just as it is" does not place all unobservables on the same footing. We'll tentatively believe some theoretical claims, yet regard others as parts of theories that are merely useful. Furthermore, our acceptance of certain theoretical claims as true shouldn't involve any metaphysical baggage. When we say that it is true that there are molecules, we shouldn't be committing ourselves to any particular account of truth (such as truth as correspondence with the world).

The NOA isn't a particularly appealing option for realist wanna-be's, however. Fine's refusal to accept even an informally realist understanding of claims to truth in science takes away whatever concession to realism seems to be suggested by his way of speaking. In addition, his view has no provisions for securing the view which is most distinctive of realism: the view that we are sometimes rationally compelled to accept certain beliefs in unobservables (not simply that such beliefs are rationally acceptable).

Fortunately, a further option is still available: retain the hope that realism can be adequately supported, but dispense with natu-ralism, and the externalist or reliabilist conceptions of how episte-mological problems should be solved that come with it. Perhaps there is still some room for a defense of realism that is transcen-dental, a defense which appeals to the very preconditions for thinking or speaking meaningfully about the world, and is based on insights into what we take to the world in order to begin inves-tigating it properly (a priori constraints), rather than insights which we take from the world after we have looked at it (a poste-riori justified rules of inference).

This kind of approach requires a dual shift in understanding: a change in what we take realism to be, and how we think it should be defended. If realism is construed naturalistically as an empirical thesis about our having epistemic access to the world ("we have approximately true knowledge of unobservables"), then the thesis of realism would seem to require empirical means of defense which are akin to the methods used in science; circularity results. If, on the other hand, we construe realism non-naturalisti-cally as a uniquely philosophical thesis about the nature of the con-straints of rationality ("we are sometimes rationally constrained to accept the best explanation as true"), we leave open the possibility that reflection on a priori or transcendental constraints on thought and meaning can secure it.

CHAPTER SIX

The Transcendental Road to Realism

Convinced that he has reduced the traditional explanationist defense to rubble, Fine moves on in "Piecemeal Realism" to see whether any of the remaining approaches to securing realism bear greater promise of success. He is quite unhappy with the first kind of alternative he considers, the "entity" realism which Hacking and Cartwright promote. Their suggestion that the acceptance of a causal explanation entails belief in the reality of the entities invoked in the explanation (though not belief in the truth of the full fledged theories concerning these entities) is readily counterbalanced by the anti-realists' claim that acceptance of an explanation needs to involve only belief that the entities appealed to are instrumentally useful. All that hands-on talk of building and using contraptions to manipulate, influence, or detect the unseen objects of science (designing colliders at CERN for smashing particles, watching the vapor trail of a charged particle in a cloud chamber, examining chromosomal material extracted during amniocentesis to determine the sex of a fetus) can be rationally reconstructed in instrumentalist terms.

Having dispensed with Hacking and Cartwright's strategy, Fine turns to what he regards as realism's last hope and final exit: contextual realism, the sort Miller promotes. This kind of realism is more modest in scope than traditional forms of explanationism, involving commitment to the idea that explanatoriness only sometimes leads to rationally compelled belief in unobservables. One should be a realist only with respect to some of the entities posited by science.

What underlies this selective realism is not a faith in the capacity for explanatoriness to be effectively linked to truth (a conviction which traditional explanationists problematically rely on), but rather a belief that we are rationally compelled to accept certain topic-specific truisms, truisms which sometimes play a vital role in securing which scientific explanation for a given phenomenon or range of phenomena is best.

Fine's interpretation of how Miller's defense of contextual realism proceeds is only partly charitable. He appropriately highlights the role of topic-specific truisms, but seems to misunderstand how and why they can create situations where doubt is unreasonable. This shortcoming is linked, very likely, to his neglect to explore the internalist's grounds for thinking that prima facie belief in certain criterial propositions is rationally compelled. It's hard to determine why the application of truistic principles to experience might sometimes leave no room for reasonable doubt if you fail to see why we are rationally constrained to believe these evidential principles in the first place.

As a consequence of this oversight, Fine's criticisms of Miller's program are in general less than compelling. Our picture, in "Piecemeal Realism," of how radically Miller's strategy departs from traditional explanationism and its problems is unsatisfying and incomplete. In what follows, we shall assess Fine's criticisms, and make an effort to address the objections to the internalist defense of realism which we considered in chapter I (objections which are somewhat related to those Fine raises):

1. The truisms dictate that certain types of experience are, in the absence of a specific reason for doubt, to be taken as a reason to believe a particular causal account of what has given rise to these experiences. What counts as a specific reason for doubt, and why is the reason for belief indicated in a truism compelling enough to demand belief in the appropriate causal account?

2. Why isn't mere pragmatic acceptance of the truisms and the consequences of applying them to experience sufficient to secure an agent's capacity to speak meaningfully about the relevant subject matter and to participate in cooperative inquiry?

3. Einstein makes use of fundamental, yet non-truistic causal principles of physics in his Brownian motion defense of the molecular hypothesis, so how can Miller's efforts to apply the internalist strategy to it be successful?

Making some headway in responding to these questions should enhance the overall appeal of the internalist approach to defending realism.

I. FINE'S CRITICISMS OF MILLER'S REALISM

According to Fine, Miller believes that doubt about the existence of certain unobservable entities is unreasonable in some contexts because, in these contexts:

1. Relevant truistic principles dictate that we have reason to believe in the existence of the entities in question.
2. There is no specific reason for doubt.
3. In general, when we have reason to believe, and have no specific reason for doubt (what Fine calls "specific doubt"), we are rationally constrained to believe.

The power of truisms to create situations where belief in specific unobservables is not just reasonable, but also rationally compelled, is grounded in a highly general principle of rationality (3) which Miller allegedly accepts.

Fine does not call the existence of truistic, topic-specific evidential principles into question, but he does take issue with Miller's appeal to 3. He finds it quite odd that Miller, the vehement "particularist" and anti-positivist, would ultimately depend so much on a highly general principle of rationality—and a dubious one at that. Fine himself is not at all convinced that the principle expresses a genuine norm of reason. He thinks that it is connected too intimately with an erroneous conception of rationality, according to which whether one ought to believe or not believe is determined by whether reasons for belief outweigh or are outweighed by reasons not to believe. Such a conception, which is too conservative about the range and variety of potential causal agents in reasonable belief and belief suspension, is undermined by hosts of ordinary cases where reason for belief is present, and reason for doubt is not, but doubt remains reasonable.

> ...we do not charge our friends with irrationality in the denial stage
> of mourning. To the contrary, we find their denial perfectly appro-
> priate and reasonable. Similarly, despite some moments of tempta-
> tion, in our better moments we do not consider our students irra-
> tional just because the excellent reasons we present, eliminating all
> their specific doubts along the way, still fail to persuade. Believing is
> a species of doing, although not an altogether voluntary one, and
> coming to believe quite reasonably requires not just good reasons
> (qua causes) and the absence of specific doubt (qua absence of inter-
> fering conditions) but it also requires the right causal mix in all the
> surrounding conditions. There are no general rules to say when the
> mix is right. (Fine 1991, p.90)

Not all cases which fall into this same category are so ordinary. Doubting the existence of molecules, for example, would not
have been unreasonable during Einstein's time, despite the observation of the phenomenon of Brownian motion, the reason for
belief in molecules supplied by the application of the relevant truism ("If a non-living thing is in constant erratic motion, that is a
reason to believe it is constantly being moved to and fro by an
external agent"), and the alleged absence of specific reason for
doubt.

Fine has several different explanations, over and above the
lack of appeal of principle 3, for why the realist conclusion Miller
desires for this case is not forthcoming. Fine wonders whether reason for doubt was or is genuinely absent: electrostatic forces could
have been considered responsible for the motion; perhaps the phenomenon could have been seen as presenting a case of genuine randomness.

Fine also questions whether the relevant truism–a truism
which is false in light of quantum theory, and would have been
rejected during Einstein's time–is genuinely productive in this case.

As a final blow against Miller's defense of realism, Fine notes
how an anti-realist could overturn the entire scheme by rejecting
the idea that we are rationally compelled to believe in the truth of
the truisms, or in the existence of the entities that truistic eviden-

tial principles suggest. Why wouldn't mere acceptance of the truisms as empirically reliable, and mere acceptance of the relevant unobservable entities as instrumentally useful be perfectly appropriate for the instrumentalist? According to Fine, Miller fails to put forward any considerations to show why the instrumentalist stance is inferior (much the same way Boyd fails to show why the instrumentalist explanation for the success of the theory-dependent methods of science is less likely to be true than the realist one). The new explanationism has, ultimately, no more going for it than the old.

Though Fine develops a few specific objections against Miller's case for realism about molecules, I would first like to address the two more general problems Fine raises for Miller's project as a whole:

I. Miller depends on an untenable principle of rationality, namely, reason to believe, in the absence of what Fine calls "specific doubt," makes doubt unreasonable.

II. Miller gives us no reason to think that an instrumentalist can't simply adopt an instrumentalist attitude towards the truisms (accepting the truisms and their consequences when applied to experience as empirically useful, yet refraining from full scale belief).

II. THE TRUE SOURCE OF UNREASONABLE DOUBT

Fine doesn't clarify what he means by "specific doubt," yet how we evaluate whether Miller actually does accept 3, let alone whether he should, depends on how we understand this vague phrase. Does Fine think that specific doubt is absent in the following cases?:

a. (epistemic laziness) The agent simply doesn't try hard enough to find potentially accessible evidence against the hypothesis she is considering.

b. (epistemic poverty) The agent knows her available resources are so limited that she cannot afford investigations which, in principle, might yield reason for doubt (evidence against the hypothesis she is considering).

c. The agent possesses only a very weak reason for belief in a hypothesis, a reason so weak that it seems entirely possible that evidence against the hypothesis could turn up later.

d. The agent is faced with evidence which, when interpreted according to her own theoretical framework, provides reason for belief in a particular hypothesis, but, when interpreted in accordance with a possible alternative framework, provides reason for belief in a different hypothesis (incompatible with the original hypothesis).

e. The agent's favored hypothesis is more appealing than any available alternatives, but accepting it still involves explanatory loose ends, incompatibility with the results of certain experiences, ad hoc additions...

If Fine thinks that specific doubt is not present in these cases, then he would have to say that 3, applied to an agent with a reason for belief in a hypothesis (and no specific doubt concerning the hypothesis), implies that the agent is rationally compelled to believe the hypothesis, even if the agent is in one or more of the states described in a-e above. But 3, under this interpretation (whereby a-e are said not to involve specific doubt) is a principle Miller patently does not accept. Miller does not think, for example, that a person in a state of epistemic laziness, with reason to believe a hypothesis and the absence of "specific doubt," would be unreasonable to refrain from full-scale belief in the hypothesis: belief in this case might even fail to be rational because the agent resists following the norms of epistemic responsibility as strenuously as she should. He also rejects the idea that any time we've tried very hard to refute a hypothesis, yet only confirming evidence is forthcoming, we are rationally compelled to believe the hypothesis. If he didn't, how could he sustain the view that we should be selective or, as Fine would say, "contextual" in our commitment to realism? If a fair to good record of supporting or compatible evidence, and the lack of undermining evidence despite active efforts to find disconfirming evidence were all it took to make belief in a theory rationally compelled, we would have to be realists about just about every reasonably long-standing scientific entity that has survived up till now.

Perhaps, however, it would be unfair to attribute to Fine the view that all of a-e are the sorts of cases to which 3 could be applied because specific reason for doubt is absent. It may simply be too uncharitable to assume, for example, that Fine wouldn't consider possessing only a very weak reason for belief as a case where specific doubt or reason for doubt is absent; he might think that the reason for belief being weak introduces reason for doubt all on its own.

With a-e eliminated as potential sources of counter-examples, 3 becomes more plausible, and much more like the sort of principle a philosopher could actually accept. We can then wonder whether, under this interpretation, Fine's reasons for thinking that 3 is false are good ones, and whether Miller genuinely depends on this general principle of rationality in the first place.

Fine's case against 3 isn't particularly strong. He argues from counter-example, yet the examples he chooses aren't well-suited to illustrate why the general principle may be flawed. It seems to me that ordinary intuition favors the verdict that a bereaved person in a state of denial, or a student who fails to be persuaded by good arguments for a position (which accommodate all her objections) are not epistemically rational. The latter example only *seems* to work if you grant that the student has reason to suspect that her teacher isn't a proper authority of the subject, or that the subject matter is intrinsically so complicated that genuine authority in it is almost impossible to attain. In either case, specific reason for doubt is present, so 3 can't apply anyway.

I suspect that many counter-examples we might try to devise to support Fine's criticisms of 3 would meet the same fate: they only seem to work to the extent to which they succeed in hiding the presence of genuine reason for doubt. The implications of specific cases are hard to establish in part because "specific doubt" is such a murky phrase. If a belief which is supported by all but not undermined by any of the available evidence fails to be rationally compelled, what accounts for the lack of rational constraint? A condition which is properly called "a reason for doubt," or some other thing? The more charitable and permissive you are in what can be legitimately described as a reason for doubt (in part by thinking that a-e do present reasons for doubt, e.g. a reason for belief being weak itself introduces a reason for doubt), the more acceptable 3 becomes—perhaps so much so that 3 should be thought of as analytically or trivially true.

I'm not sure that 3 should be taken in this way but, in the end, I don't think it matters. Miller doesn't really depend on 3 in all its generality to secure the foundation of his realism; he only needs to persuade us that reason for belief, in the absence of specific reason for doubt, sometimes makes belief rationally compelled, namely, when the reason for belief is of a particularly strong kind, the kind supplied by topic specific truisms.

We appear to have plenty of ordinary beliefs, acquired on the basis of familiar evidence (the testimony of people we know well, the reports from news agencies which we trust, statements about research procedures and results from scientific authorities...) which cannot be reasonably doubted, so it seems clear enough that sometimes reason to believe in the absence of specific reason for doubt makes belief rationally compelled. We would be hard pressed to deny that an agent who is exposed to the usual evidence, but rejects the following claims, is irrational:

- James Dean is dead.
- Fast food restaurants are not controlled by outer space aliens.
- The CIA does not transmit secret messages on "Sesame Street"

We say this despite the possibility that someone might believe some sort of radical conspiracy theory, according to which the sources of information we rely on in forming the above beliefs are undetectably deceptive. (That is part of the reason why it is so hard to shake the convictions of those who accept conspiracy theories. According to the believer, it's only natural that we shouldn't be able to find evidence of the outer space aliens' control of earthly affairs, or the CIA's involvement in PBS shows for kids: they're covering it all up!)

What makes doubt unreasonable in these cases is not simply the quantity of the evidence for the truth of these propositions, or the lack of evidence for an undermining conspiracy theory (it's built into such theories that the available evidence will seem to suggest the falsehood of the theory). Sometimes having evidence of a certain quality or character is sufficient to guarantee that doubt cannot be rational.

What underlies the unreasonableness of doubt in the claims above, and others like them, are Miller's truistic, topic-specific evidential principles (evidential principles according to which certain types of experiences must, in the absence of specific reason for doubt, be seen by the rational agent as having been caused by certain kinds of facts in the external world). These experiences must, on pain of irrationality, be taken as overwhelming or strong evidence for the truth of certain claims about the world. Rejection of the claims about the world in light of the appropriate experiences is irrational, not necessarily because reason for belief in the absence of reason for doubt in general makes belief rationally compelling, but rather because the reason for belief provided by the truistic evidential principles is of a particularly strong kind: rejection of the appropriate causal explanation is self-subverting, robbing the agent of the possibility of meaning and inquiry with respect to the relevant subject matter. In these sorts of cases, failure to believe constitutes failure to grasp the relevant concepts.

This idea is, of course, most famously developed by Wittgenstein, who claims that the very notion of inquiry itself is empty if the practice of inquiry is not constrained by fundamental, required rules or norms of evidence and justification which are not, and need not be answerable to anything else. These norms and the basic commitments which accompany them, shape meaning. Meaning, consequently, is a source of the ultimate limitations and requirements which govern rational investigation: it can dictate how we must "go on" as we amass beliefs on the basis of our experiences and previous beliefs, and serves as an important arbiter of what, among our basic experiences, counts as evidence for what, and when a judgment stands in need of justification, or should be deductively or non-deductively inferred from other claims.

If meaning does play this role, we see it at work most clearly and uncontroversially in realms which are typically thought of as non-empirical, such as logic and mathematics. When we draw legitimate conclusions using strictly a priori or analytic means, how, in the absence of being able to look to the world, can we possess good standards about which judgments and inferences are in order, and which are not? By unpacking or grasping the content of the relevant concepts. It is simply part of meaning of the truth-functional connective of conjunction, for example, that A follows from A $\wedge$ B. Anyone who would cast doubt upon whether A genuinely follows from A $\wedge$ B, or who would think that a further justification of the legitimacy of this move is required betrays a failure of understanding. The basic inference rules that involve conjunction are criterial, in Wittgenstein's sense: their acceptance is necessary for genuine comprehension of the concept of conjunction.

Significantly, Wittgenstein feels that criteria exist for empirical concepts as well. Taking certain experiences as a prima facie sign that they have a particular kind of cause is sometimes required for meaningful talk of a given subject matter, e.g. if you don't regard the sweat, tears and cries of a woman in labor as prima facie signs that she is in pain—signs that don't require a further justification to be regarded as such—you can't be taken as someone who understands what "pain" means. This view seems to have been appealing to Wittgenstein at least in part because it accords so well with our intuitions, first of what could secure the very possibility of justification and inquiry, and second of what sorts of facts and epistemic practices make our commitments genuinely cognitive and meaningful.

To draw out the intuitions in the first group, we need to question, along with Wittgenstein, whether a belief about the world (and not simply about the content of our immediate experiences) must always have a justification in order to be rational. Assuming that circular justifications are unsatisfactory for the obvious reasons, we don't want to get caught in an infinite regress by insisting that we need to justify the inference which allowed us to make the judgment by appeal to yet another principle or rule of inference which would, in turn, need to be justified by appeal to some further principle or rule of inference, etc. So mustn't something hold fast without being answerable to anything else if we are to get started on a project of sanctioning or censuring judgments? Even the project of calling beliefs into question, or casting doubts presupposes, ultimately, some fundamental rules and commitments about when judgments are in order and when they are not which are not defensible by appeal to further considerations. Again, to insist here that all the relevant rules and principles must be defensible by appeal to further considerations is to embark upon an infinite regress.

Some truistic commitments, like "the world did not begin with me," and evidential principles—such as the one which demands that I conclude that my cat Claudio is alive on the basis of my current experiences of him (I see him jump on the table, I feel him rub his forehead against my hand, I hear him purr...)—are so basic and central to our way of thinking that there really is nothing we could appeal to in the course of attempting to justify them which would strike us as more solid than the commitments and evidential principles already are. In fact, a case could be made for saying that nothing could conclusively speak against them in the first place. If we did try to reject or cast doubt upon these judgments, we would rob ourselves of the basic equipment we need to begin to play the game of distinguishing between the rational and the irrational, of seeking understanding, of trying to figure things out, of conceptualizing the world... If I demanded, for example, that every principle of causal ascription that I make use of in forming judgments about Claudio must possess a justification, I would never get anywhere. I would need to defend my taking Claudio's jumping, purring, and rubbing as a sign that he is alive by appeal to some other causal or evidential principle which in turn would need to be defended by appeal to yet another causal or evidential principle, etc. I would fall into an infinite regress, and find myself subject to the peculiar requirement that I defend a claim which is already so compelling, if anything is, that I can't even picture what sorts of considerations could be raised in its favor which might strike me as more compelling. Correspondingly, if I refused to rely on the most truistic evidential principles which are relevant to inquiry about animals, I would have nothing left to work with, and would have to give up on the project of forming judgments and trying to find out about Claudio.

Reflections on what is required for justification and inquiry to be possible have led us to Wittgenstein's conclusion that we must accept some basic evidential principles, and that we must regard these principles as perfectly in order, just as they are–no further justification is required for our reliance on them to count as rational. But what makes the particular evidential principles and truisms which inform our actual, common-sense view of the world rationally required? We have already begun to answer this question by suggesting that rejection of the basic evidential principles and truisms relevant to a particular subject matter makes inquiry into that subject matter impossible, but to defend this response more thoroughly, we need to explore the second group of intuitions informing Wittgenstein's view that there are criteria for empirical concepts, intuitions about what sorts of facts and epistemic practices make our commitments genuinely cognitive and meaningful.

Wittgenstein generally speaks of the relation criteria bear to meaning, but note that we could just as well follow Christopher Peacocke and stress the bearing acceptance of basic evidential principles has on conceptual grasp, rather than on some sort of purely semantic mastery. (Grasp of a concept is, for many of us, intimately related to grasp of the meaning of a term, but, by emphasizing the former instead of the latter, we run less of a risk of alienating someone who favors a causal or direct reference semantic theory; we don't have to commit ourselves to a particular approach to solving some of the classic problems in philosophy of language.) What intuitions are important for supporting the view that there are a host of common sense, empirical criteria, and inferences involving these criteria which we are required to accept, on pain of irrationality and failure of conceptual grasp?

Turning to our intuitions about what could be required for bona fide understanding of a concept, we see that much more must be involved than simply having the right sort of representation or image in one's mind[1], or being able to point to some things that fall under the concept, or even being able to pick out some generally accepted statements about the thing the concept is a concept of. Of course these could all be necessary conditions for comprehension, but they won't on their own be sufficient. Very young children, for example, can readily display this sort of competence without our thinking that they should be counted as having mastered the relevant concepts. A four year old who lives in a city might be very good at being able to identify drug addicts or alcoholics who live on the street, and may be able to say some true things about drug addicts or alcoholics ("these drug addicts and alcoholics are unhappy and sick, they need help, and spend a lot of money on something that is not good for them..."), but, unless the child is wise well beyond her years, we don't think her successes show that she possesses the concepts of drug addiction and alcoholism. Similarly, an adult may be able to use commonplace theoretical terms (such as "schizophrenia," "radioactivity," "electrostatic generator") quite effectively in very limited contexts, even if she fails to grasp the relevant concepts. A person can correctly point to objects that fall under a theoretical term, state claims that are true about these objects, and make good judgments for action on the basis of these claims without knowing enough about the theories in which the scientific concept is embedded to be properly said to understand it. The same holds for someone who is congenitally blind, yet displays rich linguistic competence with respect to the color terms: no amount of success at being able to say true things about blueness, or to remember and identify objects which are described as "blue" will convince us that the subject possesses the concept *blue*.

---

1. In *The Blue and Brown Books*, Wittgenstein argues persuasively that grasping a meaning, or understanding a word doesn't consist of having a certain representation of the thing the word stands for in the mental medium. Such a representation is just another sign or picture which can be interpreted in a number of different ways, as standing for a number of different things: we've simply passed the buck to a new kind of sign. The same point can be made for conceptual grasp.

If limited linguistic mastery of the sort we have just discussed is not enough, what facts, capacities or dispositions ground genuine concept possession? An attractive candidate, implicitly proposed by Wittgenstein, and actively promoted by Peacocke and Miller, is the acceptance of the relevant criteria, the possession of the disposition to infer certain claims concerning the subject matter on the basis of certain kinds of evidence or experience. (Wittgenstein and Miller would also want to include the acceptance of relevant truisms, not all of which will necessarily take an evidential form.) A concept of an object is a way of cognizing an object, and nothing is more important to determining the character or nature of a particular way of doing this than the placement of the object in a scheme of reasons, evidence, explanations, inference rules... A person who understands a concept, then, must have a tendency to reason with it in the ways which are most fundamental to its content: she must know what counts as evidence for facts about the objects which fall under the concept, what provides good reason or bad reason for forming beliefs about these objects, what allows us to move properly from one belief to the next, etc.

This knowledge must be fairly comprehensive. We want to be able to conclude, for instance, that a struggling logic student who thinks that B follows from A→B and A, but doesn't think that -A follows from A→B and -B hasn't really grasped the concept of material implication: she accepts some, but not all of the inference rules for the concept.

This particular case involves only deductive reasoning, but the same demand for thoroughness holds for empirical concepts. In all the cases of failure of conceptual grasp which we looked at above, the agents lack a comprehensive capacity to make basic appropriate inferences about the subject matter on the basis of certain standard types of evidence or other beliefs. They may be able to make some legitimate inferential moves, but the principles which they have succeeded in internalizing are far too limited and unimportant to secure concept possession. To count as grasping the concept *blue*, for example, one must have a disposition to form the judgment that an object is blue on the basis of the visual experience of that object having a certain kind of qualitative character (namely, the kind associated with objects being blue) under standard conditions. A congenitally blind person cannot make this sort of inference, and so cannot possess the concept *blue*.

Even when acceptance of the evidential principle is necessary for conceptual grasp, the requirement that certain kinds of experiences are to be taken as evidence of their having been caused by a particular sort of phenomenon might very well, in some cases, be defensible by appeal to further empirical considerations. For example, anyone who fully grasps the notion of high blood pressure must have some understanding of what sorts of mechanisms and instrumental readings serve as signs for high blood pressure, even though the fact that these signs are evidential requires further empirical support. Nevertheless, some of the inference forms or evidential principles which constitute a concept's criteria (such as the evidential principle we highlighted for the concept *blue*) will have to strike the concept possessor as not standing in need of further justification–they must, to use Peacocke's phrase, be "primitively compelling."

These sorts of evidential principles are ultimately the most fundamental and indispensable. As we saw earlier, the whole project of trying to form a conception of when a judgment about a subject matter is warranted and when it is not becomes impossible without them. Thinking that a further justification or support is required for evidential principles and inference forms in this class shows as much of a failure to grasp the relevant concepts as not accepting these evidential principles and inference forms in the first place. Someone who believes that A follows from A ∧ B, but also thinks that this inference must be supported by appeal to something else has simply failed to heed the rules of the game for the concept of conjunction. The legitimacy of this rule falls right out of the content of the concept, a concept which would be empty, or not a concept of conjunction at all without it. Someone who feels that having a visual experience with a certain sort of qualitative character under standard conditions does not on its own give compelling reason to think that the observed object is a particular color cannot properly be said to possess the color concepts. To think that something more is needed—a something more which we cannot even imagine—betrays deep misunderstanding.

Fine completely overlooks the availability of a Wittgensteinian defense of the view that truisms can make doubt unreasonable, and fails to appreciate how the introduction of a apriori considerations facilitates the formulation of a new kind of explanationist strategy for securing realism (a strategy which Boyd, with his philosophical naturalist rejection of the a priori, cannot embrace). With the exception of the objection discussed below, Fine fails to raise any criticisms which might begin to call this new explanationism into question.

III. WHY TAKING ON ISN'T GOOD ENOUGH

Fine's objection that Miller fails to demonstrate why an instrumentalist can't take an instrumentalist stance towards the topic specific truisms is part of a larger problem that can be raised for the internalist's Wittgensteinian defense of the claim that belief in truisms is rationally compelled. Why can't we be more flexible in our understanding of what kind of cognitive stance towards the truisms is sufficient to avoid self-subversion? There are any number of kinds of commitments which fall short of genuine belief, and why shouldn't simply "taking on" or pragmatically using the truisms (refraining from believing them, but keeping them in one's head and using them to guide projects, to influence actions, to make choices about which other claims should be "taken on" in light of experience...) be enough to secure the possibility of inquiry and meaning with respect to the relevant subject matter?

If we are rationally required only to take on the truisms, then we are rationally required only to take on the consequences of applying the truisms to experience. As a result, we are never rationally constrained to believe certain claims about unobservables (or to believe in the existence of any particular unobservable, even an unobservable which has a truistic justification); pragmatically taking on such claims is all that is rationally compelled. This conclusion seems at odds with realism, which we typically take to involve the idea that we are rationally compelled to *believe* in the existence of certain unobservable entities.

Is the internalist defense of realism—which relies on the claim that prima facie belief in the truisms, in the absence of specific reason for doubt, is rationally required—undermined by the suggestion that rationality and our interest in avoiding self-subversion demand only that truisms be taken on, a stance allegedly compatible with failing to believe them?

The most straightforward response to Fine's objection is to note that we have already argued for the claim that concept possession sometimes requires that an agent regard certain inferences or evidential principles as primitively compelling. To take on, yet refuse to believe a conclusion which is based on these sorts of criterial evidential principles applied to experience is, essentially, to cast doubt on the legitimacy of the move and to suggest that "something more," some further justification, would be required for full scale belief to be in order. But we have just seen how thinking that truistic, criterial inference rules and evidential principles stand in need of further justification contributes as readily to failure of conceptual grasp as not accepting the inference rules and evidential principles in the first place.

Here are some additional reasons for thinking that the internalist defense of realism is not threatened by Fine's "taking on" objection (in order of increasing attractiveness):

a. The view that we should use, but not literally believe claims about the world has a long history. The ancient Pyrrhonist skeptics, for example, thought that "appearances," the way things strike us as being, should guide our action, but not command our assent.

Though retreat to this kind of radical skepticism may seem to give an anti-realist a way to circumvent realism, quite a few anti-realists would want to refrain from pursuing this tactic. Van Fraassen, in particular, has explicitly expressed his dissatisfaction with wholesale skepticism, wanting only to cast doubt on our capacity to form a singularly rational or true picture of the unobservable part of the world—not the observable part. I assume he would want to say that at least the rudimentary truistic evidential principles (connecting our most basic experiences of mid-sized objects to their causes in the external world) have some rational force. We really do, on pain of irrationality, have to believe in the existence of people, trees, cats, etc.

These sorts of more moderate anti-realists are, in most contexts, the realist's relevant opponents, and they would tend to find the instrumentalist attitude towards the truisms unwelcome because of its fidelity to radical skepticism.

b. Let's suppose, for the moment, that—your ultimate cognitive attitude what you will—rationality dictates only that you must use or take on the truisms. How much damage would this concession do to the internalist defense of realism? Does this modification lead us directly to anti-realism, as our previous musings seemed to suggest?

Applying this modification to the internalist program, we would have to say that, at stages of scientific inquiry where defenses of unobservable entities rely ultimately on topic specific truisms (shared by all frameworks), commitment to the unobservable entities in question would have to be used or taken on, on pain of irrationality. Perhaps an anti-realist, who is typically wary of limitations placed on the range of rational alternatives open to inquirers (beyond the constraints of logical coherence and empirical adequacy), would find this conclusion as unappealingly realist as its more familiar counterpart involving full-fledged belief in the existence of the relevant unobservables.

Anti-realists are not inherently inimical to the thought that practical rationality can sometimes dictate that we should use or adopt certain theoretical claims. We may put a high premium on a set of pragmatic constraints which single out a particular theoretical option as preferable to all the others. For example, an anti-realist could say that, in light of Einstein's work on accounting for the phenomenon of Brownian motion, it would be best, for practical reasons, to adopt the molecular model of the structure of matter and use it as a foundation for further research. No full blown, equally empirically adequate continuous model is readily at hand,

and to develop one (however possible in principle) would involve enormous effort and expense, accompanied by numerous complications and ad hoc maneuvers. We couldn't even know at the outset that such a strange, ambitious project would succeed. Science must move forward, however, and since there are no realistically available alternatives, we should take on the molecular hypothesis (without necessarily believing it, unless we want to).

What singles out the molecular hypothesis as the best choice in this anti-realist version of the story are largely non-epistemic interests: our choice is the one which best conforms to our many non-epistemic desires without violating the broad epistemic constraints of logical coherence and empirical adequacy. That developing a continuous alternative would be excessively complicated, inconvenient, career threatening, etc. doesn't make such an alternative less likely to be true than the working molecular hypothesis.

When the anti-realist says that rationality sometimes dictates that we should take on a theoretical commitment, the sort of rationality in question is always practical and non-epistemic, however. For the anti-realist, as long as the possibility of developing an equally empirically adequate alternative is thought to be real (and thinking that such a possibility is real is a hallmark of anti-realism), a rational inquirer is forced to take on the molecular hypothesis relative only to her particular interests, goals, desires, pressures and circumstances. Removed from the contingent, pragmatic features of her situation, she is free to take on the continuous hypothesis and devote her resources to constructing a theory which is as quantitatively well matched with the observable world as the molecular theory, and provides "continuous matter" explanations of the various phenomena that "discreet matter" explains so well.

Perhaps, then, there is something distinctively realist about the claim that epistemic rationality sometimes settles which theoretical commitment should be taken on (leaving open whether an even stronger commitment might be required). Under this interpretation of realism, the realist, contrary to the anti-realist, insists that there are cases where an inquirer is rationally constrained to take on certain commitments about unobservables, even when we take into account only the interests she has in virtue of being a participant in the project of cooperative inquiry into the truth or truths about the subject matter.

Once we distinguish being rationally constrained to take on a commitment in the purely practical, and the epistemic sense (the former is constraint in light of particular desires and interests an agent possesses, the latter is constraint in light of only the desires and interests an agent possess in virtue of being a participant in cooperative inquiry into truths about the world), there seems to be a meaningful way to draw a distinction between anti-realism and realism—even when we're just talking about taking on or using, but not genuinely believing theoretical commitments. We can reformulate explanationism in the following way:

• We are sometimes rationally constrained to take on the best explanation (involving unobservables) over other, explanatorily inferior alternatives, where "rationally constrained" is to be understood in the epistemic sense: constrained relative only to the interests the agent has in virtue of being a participant in the social project of inquiry into truths about the world. Even when an agent is placed behind a "veil of ignorance" of all her circumstances and interests which extend beyond her desire to join in cooperative striving to find out truths about the world, there are cases where she can see that certain explanations involving unobservables stand out as best (i.e. they uniquely conform to and are recommended by the norms for cooperative inquiry into the relevant subject matter).

Anti-realism, or anti-explanationism, can perhaps be reformulated in a corresponding way as:

• We are never, from the epistemic standpoint, rationally constrained to take on the best explanation (involving unobservables) over other explanatorily inferior alternatives. An agent who is behind a veil of ignorance of all her particular circumstances and interests beyond her desire to participate in the cooperative project of striving towards the truth would never be able to choose a single explanation as better than its empirically equivalent, or equally empirically adequate rivals.

This new way of drawing the distinction between realism and anti-realism seems to highlight an important, distinctive feature of realism which is sometimes overlooked when we focus too much on the questions "Do we, or do we not have knowledge of unobservables?" or "Does science give us an approximately true picture of the world?" Realists think that epistemic rational constraint reaches beyond the limits imposed by logical consistency and empirical adequacy, and explanatoriness is somehow a factor in this extension. Whether the constraint in question dictates full fledged belief or simply "taking on" may not be so important.

Someone might wish to object to the reformulation by claiming that the norms for cooperative inquiry into a particular subject matter shouldn't be considered relevant to epistemic rationality. Couldn't we accept that a unique choice behind the veil of ignorance sometimes exists, and still insist that the all the norms which single out this choice as best, namely the norms for cooperative inquiry, should be regarded as purely pragmatic since they are not linked to likelihood of truth in the strong way that the constraints of logical consistency and empirical adequacy are? Isn't the reformulation of explanationism a concession to pragmatism, and so not really a form of realism which stands in genuine contrast to anti-realism?

Even ignoring the issue of reformulation, this objection poses a significant threat to Miller's explanationist defense of realism in general: his argument greatly depends on the claim that the norms for cooperative striving towards the truth (whether or not the truth is ever attained), and, more specifically, the norms which fall out of preconditions for concept possession (whether or not these concepts carve out reality correctly) are relevant to determining which explanation among equally empirically adequate alternatives is uniquely rationally acceptable—from an epistemic standpoint.

Fortunately, we can retaliate, I think, by casting doubt upon the view that, in order for a norm to be relevant to epistemic rationality, it must be perfectly clear that following the norm makes attaining the truth more likely, and rejecting the norm makes attaining the truth impossible, i.e. that every norm of reason must be as straightforwardly relevant to the likelihood of arriving at truth instead of falsehood as the rule that our theories should be logically consistent and empirically adequate.

Sometimes it is good enough for a norm to be relevant to epistemic rationality if everyone engaged in the project of trying to investigate a given subject matter can see, special interests and circumstances to one side, that following that norm reflects a good strategy for *trying* to get at the truth given our epistemic situation. Following a norm can be part of a good strategy for trying to get at the truth given an epistemic situation, even in circumstances (like being a brain in a vat) where following the norm doesn't succeed in leading you to the truth. This accounts for our intuition that people who seek out debate, allow their ideas to be challenged, and attempt to respond to objections are more rational, even if they fall into error, than dogmatists who have just happened to hit upon the truth.

We can rid ourselves still more completely from the need for a clearly visible link between truth and reason by insisting that a norm is relevant to epistemic rationality if it issues from the very preconditions for cognizing or inquiring into the world, even if some skeptical scenario is correct, and these preconditions have nothing to do with the way reality actually is. Just as Kant calls Hume's radical empiricism into question by adopting a transcendentalist stance, we can insist that the anti-realists have too narrow a conception of what genuinely constrains responsible thought: the preconditions for the possibility of cognizing a subject matter and cooperatively trying to understand the phenomena it encompasses are also relevant to deciding where reason can and cannot go. By stressing the role of a priori reflection on impartial, cooperative strategies for striving to understand the world (and the criterial commitments this entails), "epistemically rational" becomes—in contrast to how it is understood in externalist, naturalist accounts of rationality— an accessible, useful term of appraisal, indicative of when a belief has been arrived at in a way which is epistemically responsible, i.e. worthy of trust by others who are also engaged in the project of trying to find out about the world.

c. Ultimately, I would not recommend the instrumentalist

modification, even if it spares realism. It's much easier to see what the norms of epistemic rationality are when we are allowed to say that they are constituted by the components of good strategy for trying to get at the truth about the world or relevant subject matter (where the standards for what counts as a good strategy are just the standards everyone engaged in the project of inquiry into the world/the subject matter would want everyone to follow, given their interest in cooperatively trying to get at the truth, but not any nonepistemic interest they might have).

If intending to try to attain a grasp of the world is not said to accompany any of our most basic cognitive commitments (we are always simply taking on, but not believing) it's very hard to understand what function adopting cognitive commitments can genuinely have. What is the end that drives all this "taking on"? Given the extensive, rich use we make of our most basic cognitive commitments (to guide projects, to choose actions, to make decisions about which other cognitive commitments we should adopt...), perhaps the best solution is to insist that there is no genuine taking on the truisms which is not itself believing them.

We should not be too surprised if it turns out that there are strong parallels between concept possession and belief possession. If such parallels exist, our examination of preconditions for the former should lead to some insights about the latter. Though we stressed the role a theory of concept possession can have in determining what conditions an agent must satisfy in order to be said genuinely to grasp the relevant concepts, the theory is also of some use in deciding what constitute sufficient conditions for conceptual grasp. Perhaps this theory can also help us determine whether we ought to think that someone who takes on a commitment in the rich way described above actually meets the sufficient conditions for belief possession.

Both Peacocke and Wittgenstein develop the idea that what really determines when an agent has grasped a particular concept is her disposition or capacity to "go on" in certain ways: accepting new beliefs on the basis of certain kinds of evidence, regarding certain kinds of causal agents as explanations for certain kinds of phenomena, following particular inference rules... A similar point can presumably be made for belief possession. Having a belief is more than having a sentence or depiction of a state of affairs in one's head, or being able to utter particular linguistic expressions on the proper occasions, etc.; it is having a disposition to use the belief in reasoning in a particular way, to make further judgments on the basis of it, to take certain kinds of experiences as evidence for its truth, to make decisions about actions in light of certain desires... We can then challenge someone who thinks that there is a taking on the truisms which is not itself believing them to identify a way of "going on" here which is characteristic of genuine believers, but not of mere commitment users.

IV. SALVAGING REALISM ABOUT MOLECULES

On the basis of the considerations we have rehearsed in the last few sections, we can conclude that we are rationally compelled to believe the results of the application of evidential principles or inference rules which fall out of our ordinary concepts, the concepts which are a part of anyone's cognitive description of the world. This conclusion, taken by itself, should not come as too much of a surprise. After all, many everyday beliefs about the world (such as "The CIA does not transmit secret messages on Sesame Street") seem to have the sort of evidence in their favor which must lead to their acceptance, on pain of irrationality. Doubt in these cases is not a rational option.

Our Wittgensteinian story, it might be said, accounts for this verdict in a particularly satisfying way. Anyone who successfully grasps the relevant concepts will have to make certain judgments about what sorts of basic causal forces explain particular kinds of observed phenomena concerning the subject matter in question. Making moves or "going on" in a way which departs too radically from the constraints of reasoning about the subject matter which, in light of normal experience, issue from the relevant concepts will result in a failure to achieve something genuinely cognitive. Rejecting the standard causal explanation for the observed phenomena (or introducing an alternative causal account which does not accord with and cannot, even in part, be justified by the evidential principles and inference rules which constitute an ordinary concept's criterion) prevents an agent from being able to participate in cooperative inquiry into the subject matter. The norms of reason just are the norms which everyone engaged in the project of cooperative inquiry into a subject matter would want everyone to follow in virtue of their mutual interest in striving towards the truth, but not any other particular interest they might have; consequently, anyone who rejects the relevant criteria, cannot grasp the relevant concepts, and thereby cannot follow even the most minimal norms of cooperative inquiry will fail meet the most basic requirements for rational investigation of the subject matter.

With this result, we have all we need to establish that the existence of some of the theoretical entities which Miller feels we should be realists about cannot be reasonably doubted, namely those scientific entities, such as Leeuwenhoek's animalcules, whose introduction is justified entirely by available evidence and truistic, criterial evidential principles. To accept the Wittgensteinian story and grant that truistic considerations in ordinary settings have rational force, but deny that they can have this power in science would be completely arbitrary.

By locating a source of rational constraint which extends beyond empirical adequacy, logical consistency, and probabilistic coherence, and showing that this kind of constraint is operative in a vital way in scientific reasoning, at least some of the time, we have made considerable progress in defending important aspects of realism. Nevertheless, concerns about the strength of this victory might still persist. We have not yet shown that the realist verdict can also be appropriate when the justification for belief in the existence of a theoretical entity involves, but appears to go well beyond common-sense, truistic evidential principles. As Fine seems to suggest, it is far from clear that truisms play such a central role in the defense of the theory of molecules that reasonable doubt about Einstein's explanation of Brownian motion is effectively ruled out.

Someone could have any number of rationales for thinking

that doubt about the existence of molecules is reasonable, even though doubt in straightforward, nonscientific cases (doubt about James Dean being dead, doubt about the CIA not sending mes-sages secret messages on Sesame St.) is unreasonable. Here are what I take to be the two most convincing candidates:

1. The history of science shows how often previously accepted theoretical constructs have to be abandoned in the end. On the basis of this poor track record, doubt in the existence of any theo-retical entity seems reasonable, especially when sophisticated theo-retical considerations, and not just truistic evidential principles, are needed to try to justify the theory in which the entity is embedded.

There is no similar precedent of error in the sorts of situations highlighted by the James Dean and Sesame Street examples.

2. The molecular hypothesis could, for all we know, be quite effectively replaced by an empirically adequate alternative accord-ing to which matter could be regarded as continuous. Since Einstein's argument relies on complex theoretical considerations, not just truistic ones, this alternative might very well accord with basic, criterial evidential principles as successfully as molecular theory does.

Our judgments in the James Dean and Sesame Street cases rely only on truistic considerations, so we know that any alternative view, no matter how adequate empirically, won't be rationally acceptable.

Can an internalist argument for realism about molecules be salvaged in light of these suggestions?

Against 1

The case for the reasonableness of doubt in 1 seems to ride on the expectation that there is a fair possibility that our theoretical commitment will at some point cease to be adequate empirically or cease to cohere with other indispensable scientific results. Even most anti-realists, however, would probably not want to grant that there is much of a chance that the molecular theory will eventually have to be discarded because it fails empirically. If any anomalous evidence or theoretical conflict should arise (though there's no reason to think that it will), it is more than likely that adjustments would be made to parts of science which are much less dependent on central, intuitively compelling theoretical and experimental considerations—considerations (such as the law of inertia, or the quantitative derivations from the molecular kinetic model) which are so fundamental to the study of matter and motion that it is difficult to conceive of a physics without them. The same cannot be said for theoretical commitments such as belief in the existence of ether or phlogiston which are overtly speculative, drawing for their explanatory appeal on much more than foundational physical intuitions.

As much as the history of science presents a series of failures, it also presents a series of triumphant results which, though perhaps modified in some respects over the years, have stood the test of time. The sort of error that we find in the scientific enterprise as a whole cannot, on its own, be sufficient to ground reasonable doubt in the molecular hypothesis because it cannot be shown that the molecular theory is sufficiently like unsuccessful scientific explanations which have depended, for their explanatory appeal, on speculative or theoretical commitments at great remove from foundational principles and intuitions in physics.

Someone might worry that the case for seeing a crucial disanalogy between the molecular hypothesis and abandoned theoretical commitments is self-fulfilling or question-begging. Perhaps being grounded on principles and intuitions which are "central" or "foundational" in physics means no more than being immune from revision in a way which is ultimately arbitrary. The molecular hypothesis may be able to endure complications in a way unavailable to the theory of ether, not on account of any fundamental difference between the means with which they are justified, but simply because the molecular hypothesis and the considerations which support it will, in fact, be chosen to survive (perhaps no matter what).

Bear in mind, however, that we are not claiming that good likelihood of survival is evidence of truth, nor even that good likelihood of survival makes doubt unreasonable; we are merely saying that the failure in the history of science is not enough to ground reasonableness of doubt in the case at hand. Given the foundational, criterial character of the relevant evidential principles, an anticipation of the failure of molecular theory is no better grounded than an anticipation of failure in the James Dean and Sesame Street cases.

One could perhaps defend the stronger point that a good likelihood of survival for the molecular hypothesis might actually contribute to making doubt about the existence of molecules unreasonable, once the rationale for assigning a good likelihood is understood. This argument would require a defense of the claim that what we choose to have survive is not always arbitrary, and is sometimes inevitable, perhaps when the parts of scientific theory we hang on to fit particularly well with the fundamental intuitions about nature which serve as our starting point. Some of these intuitions are wholly common-sensical and truistic, but we may also want to include foundational principles, like Newton's law of inertia, which strike us as criterial—perhaps not criterial of pretheoretic, layman's concepts of body and motion, but of the basic concepts of modern physics.

At any rate, the burden of proof seems to have shifted to the anti-realist who is primarily motivated by the history of failure in science. The possibility of ultimately disconfirming evidence arising against the molecular hypothesis seems as remote as the possibility that our straightforward beliefs about Sesame Street will someday be disconfirmed. In both cases, our believing appears to reflect the same degree (or lack) of epistemic caution. We're in exactly the same kind of state as far as epistemic responsibility is concerned, so why is rejecting belief reasonable in the latter case, but not the former? What accounts for the different epistemological verdicts?

Against 2

One of the most crucial intuitions which motivates anti-realism is the feeling that an incompatible, equally empirically adequate alternative to any accepted theory in science can always in principle be developed. We never have compelling reason to believe in a particular theoretical entity, according to the anti-realist, because the only epistemically relevant features of current theory which speak in favor of the existence of the entity—namely, empirical adequacy and consistency—are also features which would be possessed just as well by a possible, empirically equivalent rival (in which the entity is rejected in favor of some alternative causal mechanism).

It is one thing to agree with Poincaré that incompatible, empirically equivalent rivals to any theory can always in principle be developed, no matter how awkward, inelegant, and contrived the end result; it is another to say that such alternatives would certainly have all the epistemically relevant features of a confirmed theory (since empirical adequacy and consistency allegedly exhaust the constraints of reason), or that at least one such alternative will accord with truisms and criterial evidential principles as well as current theory. We can accept the first claim while rejecting the rest.

Empirical adequacy and logical (and probabilistic) coherence are not the only features which are relevant to the rational acceptability of a theory: the constraints of rationality also include those which are set by the preconditions for participating in cooperative inquiry into a subject matter, constraints which, in part, issue from the possession conditions for the relevant concepts. Noting that an empirically adequate, consistent alternative to any given theory could always in principle be developed is not then sufficient for establishing that a *rationally* acceptable alternative to any given theory could always in principle be developed: one would need to show that such an alternative could accord with criterial truisms and evidential principles at least as well as the current theory.

Part of a response to the challenge to the internalist's case for realism about molecules that 2 presents might then be to shift the burden of proof. The anti-realist must provide us with a specific reason for thinking that a continuous model of matter could mesh as well with truistic and criterial considerations as the molecular model; otherwise, the expectation that it could is not sufficiently strong to ground reasonable doubt in the molecular hypothesis.

We can also try to shift the burden of proof in much the same way we attempted in our challenge to the anti-realist who is motivated by the history of failure of science, though this time we take into account the deeper source of empiricist doubt: the omnipresent possibility of empirically equivalent, incompatible alternatives.

Think again of the ordinary claims which most of us feel it would be irrational to reject or suspend judgment about:

- James Dean is dead.
- The CIA does not send secret messages on Sesame Street.

Such beliefs are accepted on the basis of common-sense evidential principles concerning what sorts of sources of information serve as reliable evidence for what sorts of facts in the world. Most of us would readily agree that it is not rational to suspend judgment in these cases, even though there are possible empirically adequate conspiracy theories which would involve the rejection of the relevant evidential principles (according to such theories, the sources of information we are relying on are undetectably deceptive.)

It's not at all clear that the alleged viability of alternative, empirically causal explanations for Brownian motion is relevantly different from the "availability" of empirically adequate, but unreasonable alternative explanations for the content of Sesame Street (such as the secret influence of the CIA) which would be suggested by a conspiracy theory. Given that Einstein controlled for, or ruled out, on quantitative grounds, possible causes from the standard repertoire of forces of the theory of his time, we can assume that an empirically adequate rival to the molecular hypothesis would have to rely on the ad hoc introduction of new sorts of causal forces (such as a strange field) which, much like the novel causal agent in our Sesame Street example, go well beyond what we or scientists would usually think of as influencing the sorts of objects in question. How, then, does the existence of radical, equally empirically adequate alternatives destroy rational compulsion in the molecular case (according to the anti-realist), but leave it intact in the Sesame Street case? Why does epistemic responsibility demand belief in the latter case, but not in the former? The anti-realist who is not a skeptic thinks that the two situations are different epistemically, but hasn't proposed a justification for drawing the line in this way. The burden of proof seems to be on her to produce such a justification.

This point seems especially strong when we consider theoretical claims from science (concerning, say the existence of evolution, dinosaurs, germs, cells, amoebas, genes, or continental drift) which seem most to reflect innovative, yet commonsensical reasoning applied to experience. Why would the existence of possible, equally empirically adequate rivals (positing any number of strange ad hoc causal mechanisms and force fields) eliminate rational compulsion in these cases, when similarly bizarre conspiracy theories fail to do so in the James Dean & Sesame Street cases? Isn't there a burden of proof on the anti-realist to show why the truistic underpinnings of certain unobservable claims in science fail to make belief in such claims rationally compelled, whereas similar truistic underpinnings for claims about more ordinary things like James Dean or Sesame street do create rational compulsion?

.

The Anti-Realist's Failure to Conform to Bayesian Expectations

Van Fraassen openly admits that he is not what he calls an "orthodox Bayesian": he believes that we are free to make radical changes in faith at any time and abandon our old probability function, without sacrificing our rationality. Nonetheless, he does accept that, under normal circumstances, when our basic likelihoods and rules for degree of belief revision do not change, we are rationally compelled to comply with the probability calculus. If he didn't regard the rules set by the probability calculus as constraining in this way, his use of Dutch books to highlight the incoherence or rational force of certain kinds of approaches to degree of belief assignment and revision would be completely unmotivated and unintelligible.

Peculiarly enough, there is a component of his constructive empiricism which completely fails to conform to Bayesian principles and expectations: his view on whether broad empirical scope, or the potential for broad scope (such as we might find in causally deep, explanatorily rich theories) is a pragmatic or epistemic virtue. If we look at this issue more carefully, we shall see that van Fraassen seems to subscribe to a questionable double standard, tolerating departures from Bayesian principles in himself, but criticizing such departures in others.

I. THE REALIST ACCOUNT OF BROAD EMPIRICAL SCOPE (AND ITS BAYESIAN JUSTIFICATION)

If there's any truth in Boyd's claim that only realism can truly motivate unification, it lies in the recognition that causally rich parts of a theory T1 which don't connect up with empirical evidence very directly within the theory itself or with the help of background theoretical considerations, may have the potential to hook up to observables in new ways once the theory is unified with or incorporated into another theory T.' Such potential for incorporation might give T1 further opportunity for support by evidence unavailable to rivals to T1 which fit the observable facts as well as T1 taken independently. This possibility for future support, which would exceed the support warranted for theories which are less conducive to unification (perhaps because they are less causally rich or explanatory), might be good reason to prefer T1 over rivals which are just as adequate empirically. Such a preference has a completely Bayesian justification, in fact.

Say T1...Tn are equally empirically adequate at a given stage of inquiry, t1: they all fit the observed facts e1 equally well, and so are assigned the same posterior probability value m at t1.

(I)   Ai 1≤i≤n, $p(Ti/e1) = m$

(The Bayesian translations of the English sentences which I offer here aren't appropriate in all contexts, but they're fine for this example. Note that "p" stands for the probability function the agent has before e1 happens, "p'," which I could have designated by "pe1," stands for the probability function the agent has after e1 happens, but before e2 happens, so, for example, $p(Ti/e1)=p'(Ti)$.)

Say T1 is richer causally, more deeply explanatory than T2...Tn, and hence is more conducive to unification with, or incorporation into other newer theories Ti' (T1 has more potential for broader scope than T2...Tn if the right sorts of theories come along), so T1 has a chance to boost its probability assignment in a way which is not available to T2...Tn.

If a new theory T1' is developed which, together with T1, happens to accommodate new evidence e2,

(II) $p'(e2/[T1 \& T1']) > p'(e2)$

evidence which is beyond the scope of T2...Tn and hence irrelevant to these theories,

(III)   Ai 2≤i≤n, $p'(e2/Ti) = p'(e2)$,

then the posterior probability of T1 will be greater than the posterior probability of any one of T2...Tn.

(*C)   Ai 2≤i≤n, $p'(T1/e2) > p'(Ti/e2)$.

Note that we can also assume that, given -T1', e2 is neutral with respect to T1:

(IV) $p'(e2/[T1\&-T1']) = p'(e2)$.

Let's call the values of the leftwise expressions in II and IV "h" and "j," respectively:

$p'(e2/[T1\&T1']) = h$

$p'(e2/[T1\&-T1']) = j$

So we see $p'(e2) = j$, and $h > j$ (from II and IV).

*Here's how we can derive *C from I-IV.*

First we use I, III, and Bayes' theorem to get :

$$\forall i\ 2≤i≤n\ p'(Ti/e2) = \frac{p'(e2/Ti)p'(Ti)}{p'(e2)} = \frac{p'(e2)p'(Ti)}{p'(e2)}$$
$$= p'(Ti)$$
$$= p(Ti/e1) = m$$

[$p'(Ti)=p(Ti/e1)$ because we defined p' as the probability function the agent has after e1 happens.]

Substituting with "m" in the appropriate way in *C, we see we want to show:

(*C) $p'(T1/e2) > m$

Using the theorem of total probability, the identity $p'(-T1') = 1 - p'(T1')$, and substituting with "h" and "j" where appropriate we get:

$$p'(e2/T1) = p'(e2/[T1\&T1'])p'(T1') + p'(e2/[T1\&-T1'])p'(-T1')$$
$$= p'(e2/[T1\&T1'])p'(T1') + p'(e2/[T1\&-T1'])(1-p'(T1'))$$
$$= hp'(T1') + j(1 - p'(T1'))$$
$$= j + p'(T1')(h - j)$$

We can then substitute this last expression for "$p'(e2/T1)$," after applying Bayes' theorem to determine $p'(T1/e2)$ (making the appropriate substitutions with "m" and "j" as well):

$$p'(T1/e2) \quad = \frac{p'(e2/T1)p'(T1)}{p'(e2)} = \frac{[j + p'(T1')(h - j)]m}{j}$$

$$= m + \frac{p'(T1')(h - j)m}{j}$$

We see that $p'(T1/e2) > m$ (our desired conclusion), because the rightmost expression in the formula above has a value greater than 0: m, j, and $p'(T1')$ are all greater than 0, and $h - j > 0$ since $h > j$.

The realist, as Boyd would no doubt agree, has no problem conforming to the Bayesian expectations for the case at hand. Causally richer, more explanatory theories can win out over (or be cognitively preferable to) rivals which are equally empirically adequate because more explanatory theories are more informative in their description of the workings of nature, an advantage that could prove especially relevant for unification. Such unification could help expand the scope of the type of evidence which can increase probability. Highlighting this feature of more informative, explanatory theories would seem to be an important part of defending a claim like Boyd's that realists are especially well suited for motivating and accounting for the practice of unification in science (accounting for the empirical success of the practice, as well as accounting for the impact unification has on our assessment of the comparative likelihood of scientific theories).

This line of thought may, in part, be what Miller has in mind when he says that (if my interpretation is correct), given two equally empirically adequate theories, both of which explain the evidence at hand, we may have good reason to prefer the more informative, richer explanation. If the more informative theory is false, it's easier to defeat through tests (against experience or against theoretical considerations), so if it continues to be as empirically successful as its rival, it's more likely to be true than its rival.

...if two rival hypotheses fit all the data equally well, and explain the phenomenon immediately in question, one may be preferred, all else being equal, because it conveys more explanatory information. In other words, it furnishes nontautologous answers to more explanatory questions, which the defeated hypothesis cannot answer. The maxim of superiority results from causal reasoning about the evolution of the hypothesis and the data. If it is false, the more informative kind of hypothesis is easier to defeat, so if it fares as well, its success is more apt to be a matter of truth, not luck (all else being equal). (Miller 1987, p.52)

II. VAN FRAASSEN'S UNBAYESIAN REJECTION OF BROAD SCOPE AS AN EPISTEMICALLY RELEVANT VIRTUE

I have defended the preference for T1 against T2...Tn on strictly Bayesian grounds, and hence have demonstrated one way in which greater explanatoriness or causal depth could be relevant to raising probability values relative to other equally empirically adequate theories. Given Van Fraassen's apparent willingness to count the Bayesian rules for belief revision as something a rational person must obey, provided she does not abandon her probability function (including likelihood assignments) in favor of a radically new one, you might assume that he would have to agree that greater explanatoriness can, in some circumstances, be rationally compelling grounds for assigning higher relative probability values, namely when such an assignment has a strictly Bayesian justification, and the agent's probability function and accompanying likelihood assignments fail to change in a radical way. The case we have considered seems to fall into this category, but is it so clear that anti-realists in general, and van Fraassen in particular, would or could accept our conclusions?

The suggestion that van Fraassen might or must shy away from a conclusion that has a Bayesian justification may seem somewhat implausible. Van Fraassen often appeals to formal, probabilistic considerations when he can use them to his advantage (think of this diachronic Dutch book argument against explanationism, or his defense of the claim that we must adhere to a principle of reflection), but this does not mean that his epistemological views always sit well with Bayesianism. In particular, he takes a distinctively anti-Bayesian stance on the question of how to view the status of empirical scope: is broad scope, the capacity to cover a broad spectrum of kinds of phenomena, or the potential for broad scope, a pragmatic or an epistemic virtue, i.e., is breadth of scope relevant to assessing the likelihood that a theory is true (or more likely to be true than empirically adequate alternatives), or is it relevant only to determining how useful a theory is? Could the scope of a theory ever rationally constrain us to regard the theory as more likely to be true than empirically adequate alternatives?

Consider Van Fraassen's familiar slogans "acceptance of a theory involves as belief only that it is empirically adequate" (Van Fraassen 1980, p.4) and "Science aims to give us theories which are empirically adequate; and acceptance of a theory involves as belief only that it is empirically adequate" (ibid., p. 12), which we can take to mean that we are only rationally constrained to accept the belief that a theory is empirically adequate, not that it is approximately true or more likely to be true than other empirically adequate alternatives. Empirical adequacy, not truth, is the overarching goal which informs the norms of inquiry for science. These anti-realist mottos provide some clues, but they don't help us settle the issue entirely.

Given that van Fraassen defines "empirically adequate" simply as "correctly describes what is observable" (ibid., p.4) (i.e. "fails to conflict with the observable facts"), it seems, at first glance, that scope cannot count as relevant to determining whether one theory which fails to conflict with the observable facts is more likely to be true than another theory which also fails to conflict with the observable facts. That a theory can successfully accommodate a broader range of phenomena (by, for example, being fruitfully unified with another accepted theory) would not rationally constrain us to believe that it is more likely to be true than a more limited alternative which is just as adequate empirically. Breadth of scope is not relevant to empirically adequacy, and so must clearly fall, it seems, into the category of pragmatic virtue, a strikingly non-Bayesian conclusion.

At the same time, however, we should not neglect to take into account the fact that Van Fraassen contrasts empirical adequacy *and* empirical strength with pragmatic theoretical virtue: .".. what sense can we make of theoretical virtues (such as simplicity, coherence, explanatory power) which are not reducible to empirical adequacy or empirical strength." (ibid., p.71) By drawing the contrast this way, Van Fraassen seems to imply that empirical strength, like empirical adequacy, and unlike the other standard theoretical virtues, is a not a pragmatic virtue. Does he think, perhaps, that broader scope (or suitability for unification) is part of empirical strength, and thereby an epistemic virtue- a feature of theories which is relevant to making comparative judgments of likelihood of truth?

The following passage from *The Scientific Image* shows that he does not: "If for every model M of T there is model M' of T' such that all empirical substructures of M are isomorphic to empirical substructures of M', then T is *empirically at least as strong as T'*." (ibid., p.67) If we are evaluating the relative empirical strength of two empirically adequate theories, where T' covers all the same observable phenomena as T, and then some (because T' is broader in scope than T), T will be empirically at least as strong as T', but not necessarily vice versa. Noting that the empirical substructures of a model of a theory is simply the part of the model which maps onto or represents the observable facts covered by the theory, we see that the empirical substructures of every model M of T will be isomorphic to empirical substructures of every model M' of T', since T' covers all the observable phenomena T covers, but all the empirical substructures of every model M' of T' will not be isomorphic to the empirical substructures of every model M of T, since T' covers a broader range of phenomena than T. Broad scope, then, is not a part of empirical strength. The best we can do to draw a connection between these two notions is to note that an empirically adequate theory which is narrower in scope is empirically at least as strong as an empirically adequate rival which is broader in scope.

Further evidence that van Fraassen regards broad scope as a

nonepistemic virtue appears, more straightforwardly, in a later passage where he lumps breadth of scope and suitability for unification along with other pragmatic virtues, and explicitly contrasts these with empirical adequacy and strength:

> When a theory is advocated, it is praised for many features other than empirical adequacy and strength: it is said to be mathematically elegant, simple, of great scope, complete in certain respects: also of wonderful use in unifying our account of hitherto disparate phenomena, and most of all, explanatory. Judgments of simplicity and explanatory power are the intuitive and natural vehicle for expressing our epistemic appraisal. What can an empiricist make of these other virtues which go so clearly beyond the ones he considers preeminent? (ibid., p.97)

Works Cited

Boyd, R., "On the Current Status of Scientific Realism," in Boyd, Gasper, and Trout (eds.), *The Philosophy of Science* (MIT, 1991).

Boyd, R., "Observations, Explanatory Power, and Simplicity: Toward a Non-Humean Account," in Boyd, Gasper, and Trout (eds.), *The Philosophy of Science* (MIT, 1991).

Boyd, R., "Realism, Approximate Truth, and Philosophical Method," in Papineau (ed.), *The Philosophy of Science* (Oxford, 1996).

Cartwright, N., *How the Laws of Physics Lie* (Oxford, 1983).

Christensen, D., "Clever Bookies and Coherent Beliefs," *The Philosophical Review*, vol. C, no.2 (1991).

Church, A., "On the Concept of a Random Sequence," *Bulletin of the American Mathematical Society*, 46 (1940).

Earman, J., *Bayes or Bust?* (MIT, 1992).

Fine, A., "Unnatural Attitudes: Realist and Instrumentalist Attachments to Science," *Mind*, 95 (1986).

Fine, A., "Piecemeal Realism," *Philosophical Studies*, 61 (1991).

Fine, A., "The Natural Ontological Attitude," in Papineau (ed.), *The Philosophy Science* (Oxford, 1996).

Gnedenko, B. V., *The Theory of Probability* (Chelsea, 1962).

Good, I. J., "A Causal Calculus (I & II)," *British Journal for the Philosophy of Science*, 11, 12 (1961).

Goodman, N., Fact, Fiction, and Forecast (Harvard, 1955).

Hacking, I., "Experimentation and Scientific Realism," *Philosophical Topics* 13 (1982).

Hacking, I., *Representing and Intervening* (Cambridge, 1983).

Hempel, C., *Aspects of Scientific Explanation* (Free Press, 1965).

Hempel, C., *Philosophy of Natural Science* (Prentice-Hall, 1966).

Howson, C., and Urbach, P., *Scientific Reasoning: The Bayesian Approach* (Open Court, 1993).

Hume, D., *An Enquiry Concerning Human Understanding* (Oxford, 1975).

Miller, R., "Absolute Certainty," *Mind*, (1978).

Miller, R., "The Norms of Reason," *The Philosophical Review*, Vol. 104, No. 2 (1995).

Miller, R., *Fact and Method* (Princeton, 1987).

Peacocke, C., *A Study of Concepts* (MIT, 1992).

Poincaré, H., *Science and Hypothesis* (Dover, 1952).

Putnam, H., *Reason, Truth and History* (Cambridge, 1981).

Reichenbach, H., *The Direction of Time* (University of California, 1956).

Reichenbach, H., *The Theory of Probability* (University of California, 1949).

Russell, B., *The Problems of Philosophy* (Prometheus, 1988).

Salmon, W., *The Foundations of Scientific Inference* (Pittsburgh, 1967).

Salmon, W., *Scientific Explanation and the Causal Structure of the World* (Princeton 1984).

Smart, J.J.C., *Between Science and Philosophy* (New York, 1968).

Strawson, P. F., *Introduction to Logical Theory* (Methuen, 1952).

Suppes, P., *Probabilistic Metaphysics* (Blackwell, 1984).

Suppes, P., *A Probabilistic Theory of Causality* (North-Holland, 1970).

Tarske, A., "The Semantic Conception of Truth," *Philosophy and Phenomenological Research* 4 (1943).

Teller, P., "Conditionalization and Observation," *Synthese*, vol. 26 (1973).

van Fraassen, B., *The Scientific Image* (Oxford, 1980).

van Fraassen, B., "Belief and the Will," *Journal of Philosophy* 81 (1984).

van Fraassen, B., "The Peculiar Effects of Love and Desire," in
 A. Rorty and B. Mclaughlin (eds.), *Perspectives on Self-
 Deception* (University of California, 1988).
van Fraassen, B., *Laws and Symmetry* (Oxford, 1989).
von Mises, R., *Probability, Statistics and Truth* (Dover, 1981).
Wittgenstein, L., *The Blue and Brown Books* (Oxford, 1958).
Wittgenstein, L., *On Certainty* (Blackwell, 1969).

Index

adequacy, empirical, 2, 4, 16, 68,
  70–71, 141, 145, 149–150,
  152, 181, 188–191, 194, 198
anti-realism, 1–4, 21, 37, 143,
  147–152, 157, 178–179, 181,
  188–191, 193–200

Bayes' theorem, 8, 85–87, 91–94,
  102, 195
Berkeley, G., 69
Borges, J. L., 15
Boyd, R., 11–14, 35, 36, 38,
  75–76, 79, 81, 125–128,
  133–156, 193, 196
Brownian motion, 30–33,
  133–134, 178–179, 185–186,
  190

Cartwright, N., 36, 161
causation:
  concept of, 17
  statistical-relevance model
    of, 16
CERN, 161
Christensen, D., 108–109
Church, A., 54, 59
coherence, probabilistic, 68,
  95–103
concepts, 74, 169–176, 183–5

confirmation, 19–21

Darwin, C., 140
Descartes, R., 69, 127
Dutch books, 82–85, 89, 91,
  95–124, 193

Earman, J., 39, 67, 97
Einstein, A., 30–33, 133–134,
  178–179, 185–186, 190
empiricism, constructive, 67–71
equivalence, empirical, 67,
  145–147
explanation:
  covering-law model of, 16,
    17–18
  deductive-nomological
    model of, 16, 37, 40–41
  inductive-statistical model
    of, 17, 37, 41–44
  inference to the best, 2–11,
    77–81, 89, 133–136,
    146–147
  statistical-relevance model
    of, 45–65
externalism, 11–12, 71, 74,
  79–80, 81, 93, 126–128,
  154–155, 159

Falwell, J., 29
Feyerabend, P., 152
Fine, A., 14, 125, 128, 130,
    136–138, 157–158, 161–167,
    176–177, 185

Good, I., 36
Goodman, N., 127, 149

Hacking, I., 161
Hempel, C., 36, 40–45, 49, 51
Howson, C., 98–101
homogeneity, objective, 50–53
Hume, D., 16, 36, 38–40, 45,
    127, 148, 153–156, 182
Huygens, C., 145

incoherence, probabilistic, 82
internalism, 12–14, 34, 71, 74–75,
    80, 93, 127, 155, 162,
    177–191

Kant, I., 182

Leeuwenhoek, A., 22–23
Lewis, D., 100

Mack, J., 29
M'Naughten rules, 63
Miller, R., 5–6, 11–34, 35, 38,
    161–169, 174, 176, 181, 185,
    196

naturalism, 12, 71, 79, 126–128,
    155–157, 159
Newton, I., 139, 145

Peacocke, C., 172, 174, 183
Poincaré, H., 145, 189
pragmatism, 69–70
principle of reflection, 95–6,
    103–115
probability, subjective, 97
propositions, hinge, 15, 24–27
Putnam, H., 69

Pyrrhonism, 177

randomness, mathematical, 54–59
rationality, 1–4, 23–29, 95–103,
    114–115, 167–191, 193
realism:
    abductive argument for,
        75–76, 128–148, 150
    about molecules, 30–33,
        184–191
    contextual, 161–169
    defined, 4
    versions of, 5–11
Reichenbach, H., 36, 73
reliabilism, 126, 153–154, 156,
    159
Russell, B., 5

Salmon, N., 36, 45–65
skepticism, 24, 69
Smart, J., 72–73, 75, 128–133,
    137
Strawson, P., 155
Suppes, P., 36

Tarski, A., 138
Teller, P., 100
truisms, topic-specific, 14–16,
    21–27, 29–33, 162–165, 169

unobservables, 4, 21–23, 35, 68,
    158, 163, 176, 180
Urbach, P., 98–101

van Fraassen, B., 1, 4, 8, 9, 13,
    16, 67–124, 129, 152, 177,
    193–200
von Mises, L., 96

Wittgenstein, L., 15, 24–27, 30,
    169–176, 183–184